Herick O. Otieno
Joseph L. Awange
Energy Resources in East Africa
Opportunities and Challenges

Herick O. Otieno
Joseph L. Awange

Energy Resources in East Africa

Opportunities and Challenges

with 52 Figures and 19 Tables

 Springer

Prof. Herick O. Otieno
Maseno University
Department of Physics
Private Bag
Maseno
Kenya

Email: othienoho@swiftkisumu.com

Prof. Dr.-Ing. Joseph L. Awange
Department of Spatial Sciences
Division of Resources and Environment
Curtin University of Technology
GPO Box U1987
Perth, WA 6845
Australia

Email: jawange@yahoo.co.uk

Library of Congress Control Number: 2006928829

ISBN-10   3-540-35666-5 Springer Berlin Heidelberg New York
ISBN-13   978-3-540-35666-0 Springer Berlin Heidelberg New York

Springer is a part of Springer Science+Business Media
springer.com
© Springer-Verlag Berlin Heidelberg 2006
Printed in The Netherlands

Cover design: E. Kirchner, Heidelberg
Production: Almas Schimmel
Typesetting: camera-ready by authors
Printing: Krips bv, Meppel
Binding: Stürtz AG, Würzburg

Printed on acid-free paper   30/3141/as 5 4 3 2 1 0

The writing of this book could not have succeeded without some sacrifice. This book is dedicated to our wives Mrs. Beatrice Othieno and Mrs. Naomi Awange whose family time was sacrificed.

Herick O. Otieno and Joseph L. Awange
March 2006

Preface

Assessment of commercial energy sources such as oil products and electricity generation is a straight-forward task. For oil-based fuels, all products enter into the region through known points and are kept in centralized storage facilities where the quantity and quality can be assessed. Consumption can also be accurately determined from the records of daily collections. Similarly, electricity production is easily obtained from generation records. On the other hand, assessment of consumption of biomass fuels is very complicated and in all cases only estimated quantities can be obtained. This is due to the fact that production or re-generation takes a long time and is done by a very large number of people, organizations or institutions, each working independently.

The marketing of biomass fuels is also sporadic with highly variant prices. In addition, most of the biomass fuels used by rural households is not obtained through the market but is collected by individuals from different sources, which cannot be reasonably monitored. The figures and percentages presented in this book in reference to various energy sources are therefore averages obtained from a large number of data generated through surveys and research work done under the supervision of the authors which includes personal communications with relevant knowledgeable employees in the energy sector. A high degree of inconsistency was experienced even in cases where information was obtained from the same source at different times. For these reasons, it is most unlikely that the figures quoted here will be the same as those obtained from other sources, particularly data on renewable energies. However, for commercial fuels, the differences should not be significant.

The authors recognize the inevitable repetitions in various sections and chapters and this is because discussions on energy can be approached from various viewpoints. For example, a number of issues that affect consumption may be discussed under distribution, cost, production limitations, management policies, traditional living conditions, poverty and land ownership rights. It is therefore difficult to discuss one aspect without mentioning the other but every effort has been made to make sure that these repetitions are minimized. The other point is that some statements on energy resource development may appear to be contradictory due to differences in short and long-term energy mix preferences. Thus, as much as biomass consumption is inevitable and associated with environmental degradation, its regeneration must be supported but, at the same time, promoting a gradual switch to other alternatives, one of which is kerosene, should encourage reduction of its consumption. On the other hand, consumption of fossil fuels should also be kept at a low level for both environmental and economic reasons. So the seemingly contradictory statements and recommendations should be understood in these contexts. We would also like to note that the sequence of the chapters and sections of the book may not be in the preferred order for some readers but again this was not an easy decision as our reviewers also could not unanimously agree on the order. This shows how complex and intertwined energy issues can be.

For readers unfamiliar with East Africa, Chap. 1 provides some basic background. This is followed be an elaborate discussion on energy management in region. From availability to affordability, factors that govern energy choices are presented and the situation compared to that of the global case. Chapter 2 looks at the energy resources in East Africa and a highlight is given to those resources that are available locally and those which are imported. The potential of each country in-light of these resources are considered. In Chap. 3, the potentials of the available energy resources are outlined, the present level of exploitation given and the possible initiative to improve their application presented. Chapter 4 attempts to identify problems and challenges associated with energy development in the region. Chapter 5 is the largest in the book and attempts to highlight both technical and economic issues that must be considered in developing renewable energy. Chapter 6 presents the status of renewable energy technologies in the region while Chap. 7 identifies options and challenges regarding energy development policies.

The last Chapter gives some guidelines and suggestion on how available energy resources can be economically developed.

We take full responsibility for the accuracy of the information in this book and hope that some of our arguments will stimulate constructive debate on energy development and management policies in East Africa.

Kisumu (Kenya) and Perth (Australia)   *Herick O. Otieno*
August 2006   *Joseph L. Awange*

Acknowledgement

In the process of gathering and synthesizing materials for this book, the authors received assistance from several individuals and institutions. In this regard we would like to express our sincere gratitude to our employer, Maseno University, for providing the support and conducive working environment during the preparation of the manuscript. The first author wishes to thank the colleagues in the Department of Physics and Materials Science at Maseno University, who despite their own heavy teaching loads, had to take up his teaching assignments while he was away on sabbatical leave. The leave was granted on the understanding that it would be used to complete the book and was spent at the Department of Applied Physics at Papua New Guinea University of technology from February to December 2005. The facilities and moral support received at the Papua New Guinea University of Technology is highly appreciated.

All other friends and collaborators in Kenya, Uganda and Tanzania who contributed in one way or another are gratefully acknowledged. Special thanks go to Dr. Judith Miguda Attyang of the Department of English, Maseno University, for proof-reading the Manuscript. The first author also would like to thank his wife Beatrice Othieno and their three children Judith, Nicholas and Walter, all of whom bravely bore his absence while he was away writing the book.

The second author is grateful to Curtin University of Technology (Australia) for offering him a Research Fellowship position at the Department of Spatial Sciences, Division of Resource and Environmental. To Prof. Will Featherstone and his team at the Department, I say thanks for the warm welcome and for providing a conducive working environment and facilities that enabled the completion of the final version of the book. The author wishes to thank his wife Mrs. Naomi Awange and their two daughters Lucy and Ruth who always brightened him up with their cheerful faces. All your support, especially family time that you were denied in-order to prepare this book is greatly acknowledged.

Many people were involved in the collection of the materials that have been presented in this book but it is not possible to mention all of them in this limited space. We wish to acknowledge their valuable contributions.

Contents

1

East Africa

1.1 Introductory Remarks

When talking about East Africa, traditionally this refers to the states of Kenya, Uganda and Tanzania with a combined population of over 90 million people. The three countries have a lot in common in terms of their peoples and cultures which are dominated by two main ethnic groups of Bantu and Nilotic origins. Both groups are present in all the three states and have, over the years, interacted almost freely across the borders in line with the African traditional commitment to extended family ties including inter-marriage and many years of historical trading interactions. The same colonial government separately ruled the three states but the ties between them were so entrenched that when these countries became independent, they readily formed the East African Community which was expected to eventually evolve into one political federation. Under the Community, several important services were jointly managed and coordinated by the East African Assembly whose headquarters had been carefully planned and built at Arusha, Tanzania, complete with all the necessary secretariat facilities such as conference halls and accommodation for staff in anticipation of an eventual political union. However this expectation was not realized and, instead, the community disintegrated in 1977 but was later reconstituted as will be discussed later. Such changes in institutional arrangements and internal political instability as well as deliberately designed external factors have adversely affected the development of the energy sector in East Africa. Further, the political leaderships in the developing countries, in general, do not seem to understand the important role of energy in development process. The development of energy resources is low-keyed even though it is the driving force in national development.

In East Africa, like in many other developing countries, energy supply and resources are divided into two main groups namely commercial and non-commercial (informal) energy resources. The commercial energies are electricity and petroleum-based fuels and these are the forms of energy that usually draw the attention of national governments because their availability and distribution have strong impact on the economy and directly affect the rich and the powerful members of the society. The other group of non-commercial energies includes biomass materials such as wood, charcoal, plant and animal wastes. Although some of these especially charcoal and, to a lesser extent, wood, are today available in the local markets, they are still largely obtained free of charge by the users. Charcoal however has gradually moved from household production to dispersed commercial production and is now established as an income generation activity. It is nevertheless considered as a non-commercial energy because its production and marketing is still done on an ad hoc manner by individuals who may not necessarily be professional charcoal producers and who often switch from charcoal trade to other activities. In terms of the total energy consumption in East Africa, non-commercial energies contribute well over 80% of the total energy. Commercial energies on which the governments spend large portions of their Gross Domestic Product (GDP) contribute less than 20% of the total requirement. Non-commercial energies are not clean and pose a lot of health hazards to the users in addition to environmental degradation associated with their use. The responsibility of obtaining non commercial energies is traditionally left in the hands of women and girls who normally spend a lot of time fetching it from distant locations, and this has severely limited the participation of women and girls in other economic activities including education, making them the automatic disadvantaged groups in the society. One of the desires of the modern energy policy is to make people shift from traditional non-commercial energy resources to more modern, clean and convenient energy sources such as electricity and LPG for domestic applications. Experience in East Africa has however indicated that this is wishful thinking and will remain so unless deliberate efforts and even national sacrifices are made to give people clean energy at affordable cost. But this also requires that the quality of housing will have to be significantly improved in the rural areas. Many international studies have concluded that for many years to come, a large proportion of people will continue to rely on biomass energy on account of its availability and possibility of regeneration. The economic development of Kenya, for example, is considered to be way ahead of those of Tanzania and Uganda and yet the reliance on biomass fuel continues to steadily increase in Kenya. Another aspect that also does not seem to worry the governments is that biomass resources are obtained from standing stocks instead of from yields. This has been going on for a long time and already hardships associated with its shortages are experienced in many parts of the rural areas. The shortage also has its own implications on land quality and ultimately on the welfare of the people. So it is important to accelerate transition to modern fuels as this would reduce pressure on the land and forestry and hence make it possible to sustain good soil quality for food production and general welfare of the people in terms of poverty reduction. Although some studies have suggested a steady and gradual upgrading of energy use, for example, from biomass to Kerosene to LPG and electricity, the rule cannot be strictly applied in all situations. Energy upgrading may not be achieved through this ladder-climbing model because its optimum use for a particular user is more of the proper mix rather that quality differences.

In all the three East African states, the desire in the energy sector is to make available to the domestic economy and the people in general a cost-effective, affordable and adequate quality of energy services on a sustainable basis. The governments recognize that the success of socio-economic and industrial developments will, to a large extent, depend on the performance of the energy sector particularly the development and diversification of energy sources. Although the role of energy in development has always been crucial, very little support has been given to energy initiatives and therefore there are numerous problems that will have to be addressed in order to have a meaningful foundation for energy supply in the region. The major challenge is how to deal with the weak generation, transmission and distribution infrastructure and the policies, which have inhibited investments in the energy sector. These weaknesses and the high cost of energy are viewed as hostile conditions not just for investment but also for ordinary power consumers and, as a result, there is very low per capita power consumption in the region. Most people are forced to use own crude sources of energy such as biomass-based fuels. In view of these factors, energy resources that are locally available in the region such as biomass, solar, wind and small hydro schemes will be discussed in details including their conversion technologies. This is done in recognition of the fact that future self-sufficiency in energy for the region may significantly depend on these renewable resources and the fact that there is a general global shift from fossil-based fuel to renewable energies. The other resources will also be discussed but without the detailed conversion technologies.

1.2 Background Facts of East Africa

To be able to clearly understand the energy situation in East Africa, it is important to know certain basic important facts regarding the geography, available natural resources and the economic activities in the region. East Africa was historically known to comprise of four states: Kenya, Uganda, Tanganyika and Zanzibar. The union of Tanganyika and Zanzibar in 1964 into what is now known as the United Republic of Tanzania reduced the number of East African states to three. Both Kenya and Tanzania have Indian Ocean coastline while Uganda is landlocked.

Lake Victoria, the world's second largest fresh water lake, lies at the common border of the three states and so control of the lake's resources especially fisheries has often caused some conflict. Luckily the situation has never reached a level that would warrant international mediation. Prior to 1961, all the three states were under control and administration of the United Kingdom. Tanganyika was initially controlled by Germany but, after the Second World War, the United Nations mandated the United Kingdom to administer it. Thus all the three East African states got their independence at different times from United Kingdom. Around the time of independence, the natives of the region had a strong desire to form one federal state of East Africa but this did not happen partly due to the fact that independence was granted at different times, allowing political leaders of each state to develop their own selfish ambitions. However, the common heritage and close socio-economic relationships of the natives persuaded the leaders to establish an East African Community with its headquarters in Arusha, Tanzania. The structure of the Community had some very weak elements of federalism in which some services such as posts, telecommunication, power generation, transmission and distribution, harbours, railways, universities, high school examination system etc were jointly controlled by the three states through East African Community. The more sensitive areas like security, economic management, political leadership and crucial executive powers were left in the hands of individual state. The Community was therefore too weak and vulnerable to stand the test of time. It was finally disbanded in 1977 when the Ugandan dictator, Idi Amin could not get along with the other two leaders particularly with Julius Nyerere of Tanzania.

Kenya and Tanzania have been relatively peaceful and stable politically but there was some hostility between them due to the differences in political ideologies. Uganda, on the other hand, has had some very turbulent moments. The dictatorial regime of Idi Amin claimed the lives of about 300,000 people between 1971 and 1979. The guerilla wars and the human rights abuses under other leaders were responsible for the deaths of close to another 100,000 people. Uganda returned to relatively peaceful path of development in 1986 when Mr. Yoweri Kaguta Museveni seized power. In November 1999, the three states, with a clearer understanding of the reasons for the collapse of the original East African Community, again signed a treaty to re-establish it. At this point, it became apparently clear that there were external forces that planned and executed the fall of the old East African Community and so the leaders became even more committed to its reconstitution. The treaty came into force in July 2000 and the new East African Cooperation was established. One of the immediate tasks for the new Community was to identify areas of common economic interests that would serve as the nuclear centers for regional economic growth. For a number of good and legitimate reasons, Lake Victoria and its basin was identified and designated the pioneer economic growth zone that would foster regional socio-economic cohesion. Every state is expected to pay special attention to Lake Victoria basin with a view to promoting investment in the area in order to transform it into a real economic growth zone for East Africa. To achieve this, it is envisaged that a broad partnership of local communities, riparian states and the East African Community as well as development partners will have to be strengthened through programmes that focus on issues concerning Lake Victoria basin. It is expected that the programmes would include investment opportunities, one of which would be energy development. All these would place strong emphasis on poverty eradication, participation of the local communities and sustainable development practices that are cognizant of environmental protection. Given the important role of energy in development, it has been earmarked as one of the top priorities for the Lake Victoria region. It must be pointed out here that all the three East African partner states each has its own energy master plan that adequately covers their respective sections of the lake basin and provides massive investment opportunities in Renewable Energy Developments. The specific opportunity areas include but not limited to: energy production and supply; improvement of infrastructure to enhance distribution of petroleum products; biomass development; production and sale of biomass fuel processing kilns and stoves. In order to provide a well-coordinated management of investment issues, the Community has, in accordance with the Protocol for Sustainable Development, established Lake Victoria Basin Commission as an overall institution to handle development issues in the basin. The Commission, which is residing in Kisumu - a Kenyan city and port on the shores of Lake Victoria, recognizes the importance of participatory approach to development programmes and is ready to mobilize support for programmes initiated by the local communities and national governments. It appears this basin will in future play a very important role as the converging point or hub of the development of East Africa. Although it has a vast hydro energy potential and valuable fishery, it is also one of the areas with the highest poverty and lowest electrification levels in East Africa. It is therefore important to understand the conditions in the basin, its resources, energy potentials and the role it can play in accelerating the development of the region.

One of the most prominent natural resources that have been shared by the peoples of East Africa since time immemorial is Lake Victoria, which lies at the heart of the region. The economic activities around this lake are to a large extent a measure of what is taking place in the whole region. The area has a whole range of activities from agriculture to industries and has people of all classes from the richest to the very poor by any standards. Lake Victoria basin is also the home of three cities: Kampala the capital city of Uganda, Mwanza the second largest city in Tanzania and Kisumu the third largest city in Kenya. For this reason it is necessary to give a brief background information of the conditions in this region in order to understand energy distribution problems in East Africa. For readers interested in detailed information on Lake Victoria, we refer to [4, 5, 7, 36, 35].

Lake Victoria with a surface area of 68,800 km^2 and catchments area of about 181,000 km^2, is the world's second largest freshwater body (second only to Lake Superior of North America in size), and the largest in the developing world. It has a shoreline of approximately 3,500 km long with 550 km in Kenya, 1150 km in Uganda and 1750 km in Tanzania. The lake lies across the Equator in its northern reaches, and is between latitude 0.7°N - 3°S and longitude 31.8°E - 34.8°E. It is a relatively shallow lake with an average depth of 40 meters and a maximum depth of 80 meters. The area usually referred to as the lake basin is larger than the catchments area and is estimated to be about 193,000 km^2 distributed in Tanzania (44%), Kenya (22%), Uganda (16%), Rwanda (11%) and Burundi (7%). This area supports a population of about 30 million people, that is, 30% of the total population of East Africa lives in this area, which is only about 10% of the total area of the region. The lake itself is however shared by only three riparian states of Kenya, Uganda and Tanzania each controlling 4,113 km^2, 31001 km^2 and 33,756 km^2 respectively (Table 1.1).

Fishery is an important resource of the lake and is one of the most prolific and productive inland fisheries in Africa. At one time the lake was home to over 5,000 endemic fish species but, as pollution load increased as a result of increased industrial activities and a growing number of human settlements around the lake, the number of fish species drastically declined. There is also local belief, without any credible proof, that the introduction of exotic species particularly Nile Perch also caused the decline of fish species in the lake. Despite these changes, fisheries remain a very important source of foreign exchange earnings with an annual landed quantity of about 500,000 metric tones with an estimated value of US$ 300-400 millions. However, there are indications that if the lake is not protected against continued pollution and over fishing, then the economic value of the lake will significantly diminish.

Table 1.1. Some statistics on Lake Victoria and its Basin

Country	Surface Area (km^2)	%	Catchment Area (km^2)	%	Shoreline (Km)	%
Kenya	4,113	6	38,913	21.5	550	17
Uganda	31,001	45	28,857	15.9	1150	33
Tanzania	33,756	49	79,750	44	1750	50
Burundi	-	-	13,060	7.2	-	-
Rwanda	-	-	20,550	11.4	-	-
Total	68,870	100	181,130	100	3450	100

Over the years, dynamic fluctuations have been observed in fish production from Lake Victoria due to a number of factors but in recent years these changes began to take a generally consistent reduction trend. In Uganda, yields rose from 100,000 metric tonnes in 1980 to 132,400 metric tonnes in the 1989. From that period the catches have been gradually reducing so that in 1995 the catch was down to 106,000 metric tones and continues to fall. In Tanzania, annual catch increased from 146,000 metric tons in 1988 to 231,600 metric tonnes in 1990 and a general decline observed thereafter. In Kenya, the catch rose marginally from 186,000 metric tonnes in 1989 to 190,000 metric tonnes in 1993. Since then the catch has been going down. These trends can be explained by the fact that the apparent increases of the 1980s led to the establishment of more large scale fish processing plants along the shores of Lake Victoria, which target the international fish markets of Europe and Asia. This development raised the demand for fish and consequently led to over-fishing, which initially could not be controlled due to corruption involving powerful fish processors. The number of fish processing industries in the Region appears to have stabilized at about 34 factories almost equally distributed among the three riparian states: 12 in Kenya, 10 in Uganda and 12 in Tanzania. The irony is that although the lake's basin is rich in resources particularly fisheries, the community living in the region is one of the poorest in East Africa especially in Kenya and Tanzania. It is estimated that over 50% of them live below poverty line. Despite the vast economic potential, investment in the Basin by both local and international entrepreneurs is still very low hence high level of poverty. The relationship between poverty and environmental degradation has been demonstrated by a number of scholars and the lake basin appears to be caught up in this situation in which one condition causes the other and vice versa - a vicious circle that has caused severe damage in the region. There have been major environmental shocks as a result of deforestation, destruction and drainage of wetlands, poor agricultural practices and direct discharge of untreated (or inadequately treated) industrial and municipal wastes into the lake. Deforestation is due to a number of biomass uses but it is generally accepted that energy application is a major contributing factor.

Of the various uses/benefits from Lake Victoria, it is fish that receives most attention. Most of the fish fauna of the lake, other than the recent introductions of exotic species, lived between two million and ten thousand years ago in the west flowing rivers that later flooded to form the lake. The lake has since experienced explosive speciation particularly amongst the haplochromine cichlids, estimated to comprise over

300 species. This burst of speciation has been in response to the change from river to lake conditions. Although similar phenomena happened in other lakes, in Lake Victoria it happened much more recently, more rapidly and with fewer opportunities for ecological isolation in different types of habitats.

Within the period 1960-1990s, the native Tilapia (*Oreochromis esculentus*), previously a fish of the greatest commercial importance, virtually disappeared from the lake, but is still found in small quantities within the satellite lakes. Other fish species that have declined drastically include the migratory species, and the haplochromine cichlids. At the same time both Nile perch (*Lates niloticus*) and Nile tilapia (*Oreochromis niloticus*) have established themselves in the lake to the extent that they now dominate the commercial fisheries of Lake Victoria. The sardine-like native fish (*Rastrineobola argentea*), locally known as "*dagaa/omena/mukene*", now features prominently among the commercial catches.

The catchments area of Lake Victoria is slowly being degraded due to deforestation. The increase in human population in the riparian area has put pressure on the forests for agricultural land, timber, firewood (biomass energy) and habitation. This deforestation, coupled with bad agricultural practices, has degraded the soil leading to siltation along the rivers into the lake. Agro-chemicals and industrial effluents are now polluting the lake, while deforestation, soil erosion, and increasing human and livestock populations have all contributed to increased nutrient loading because of changing land use patterns. Sewage effluents from urban centers, beach settlements and fisher communities around the lake also contribute to the big nutrient load, which in turn has brought about eutrophication. Eutrophication has increased algal populations, caused deoxygenation of deep water and created conditions favourable for the growth of noxious weeds such as water hyacinth.

Wetlands, which normally filter the water before entering the lake, are under stress. Wetlands are reclaimed for agriculture, industrial development and human settlements, while others are drained to control human disease vectors. Some are excessively harvested for making mats, baskets and chairs. Many of the wetlands have received too much pollution to the extent that they cannot perform their filtration function efficiently. Therefore, pollutants normally retained by wetlands enter the lake unchecked, thus further contributing to the deterioration of the lake water. These wetlands also served as breeding ground for many fish species and their destruction must have contributed to the dwindling number of fish species.

Poverty in the region is rampant and, looking at the statistical figures on the Kenyan side, the trend has been upward. For instance, food poverty in Nyanza (Kenyan L.Victoria area) was way above average at 58% in 1997. Food poverty in Kisumu district rose from 44% in 1994 to 53% in 1997, while overall poverty in the city jumped from 48% in 1994 to 64% in 1997. To date, the situation has not improved.

The problems can be looked at in two categories, i.e., at micro and macro levels. At micro level, the problems are people-based and therefore, through community structures, everyone should play a positive role in improving the conditions. While at macro level the problems cut across several sectors, and the governments of the three East African countries should play a leading role by providing an enabling environment. Table 1.2[1] gives a summary of the important facts about the region while Fig. 1.2 shows the location of the three states of East Africa.

1.3 Energy in East Africa

The word energy is often used to define the level of activities. When a person is ill and is unable to talk or walk, we say the person is weak and has no energy to either walk or talk. When our soccer team is not producing the spectacular moves towards the end of the game, we say they have run out of energy. A physicist, in a simple way, would define energy as the ability to do work. However, whatever meaning we want to give energy, it is evidently clear that energy is what makes things happen and therefore work and energy cannot be separated. This being the case, energy would appear to be the prime mover of all activities. One cannot talk about development without identifying the source of energy that would provide that development. Energy is the agent for changing the state of any system; from poverty to wealth, from weak economy to strong economy, from nothing to productivity, from insecurity to safety and so on. Per capita energy consumption is therefore often and rightly used as one of the most important indices for measuring the level of development of a nation.

We learned from physics that energy can neither be created nor destroyed; it can only be transformed from one form to another. Since its

---

[1]Numerical data in the Table are averages calculated from several sources

Table 1.2. Basic Facts about East Africa

	Kenya	Uganda	Tanzania
Land area (km^2)	569,250	199,710	886,037
Water area (km^2)	13,400	36,330	59,050
Arable land (%)	8	-	4.5
Independence(year)	1963	1962	1961
Former administrator	United Kingdom	United Kingdom	United Kingdom on behalf of UN
2004 Population (millions)	30.4	26.5	36.6
Population below (%) poverty line	56	35	36
Major natural resources	Soda ash, fluorspar, limestone, some gold	Copper, cobalt, salt, limestone	Tin, phosphates, iron ore, coal, diamond, gold, nickel, natural gas
Main agricultural products	Coffee, tea, sugar cane, pyrethrum, fisheries	Coffee, tea, cotton, tobacco, fisheries	Coffee, sisal, cotton, tea, pyrethrum, cloves, tobacco, cashew nuts
Geography	Lake Victoria, Indian ocean coastline, mt.Kenya, great Rift Valley some, small lakes	Lake Victoria, river Nile, Great Rift Valley, some small lakes	Mt. Kilimanjaro, lake Victoria, lake Tanganyika, lake Nyasa, Great Rift Valley, few small lakes, Indian ocean coastline
Climate	Ranges from tropical to arid, two rainy and dry seasons per year	Ranges from tropical to semi-arid in North Eastern parts	Ranges from tropical to temperate in the highlands
Paved roads (km)	8,900	2,000	3,700
Fuel pipeline (km)	483 (petroleum)	-	29 (gas, to be extended to) 866 (petroleum)
Natural hazards	Floods, drought	-	-
Environmental issues	Water pollution, deforestation, soil erosion, poaching	Deforestation, soil erosion, poaching, wetland destruction, overgrazing	Deforestation, soil degradation, destruction of coral reefs, poaching

Fig. 1.1. Map of Lake Victoria and its Basin

consumption is a measure of development, it means that every existing
nation of the world has developed to some level and has the potential
to develop farther if it can efficiently transform available energy sources
into usable energy forms and consume them wisely. The problem with
energy however is that its high quality sources are very scarce while
the low quality sources are abundant. For example biomass, solar and
wind energies are almost everywhere on earth but they provide low
quality energy and therefore their conversion into high quality energy
such as electricity is generally expensive. On the other hand, oil and
nuclear energy sources that provide relatively high quality energy are
not widely distributed. Thus energy conversion from low grade to high
grade is a major factor in determining the choice of an energy source. If
the end use does not require sophisticated conversion techniques, then
availability becomes an important factor. In this regard, the choice for
an industrial energy use and domestic energy needs would generally

Fig. 1.2. Map of East Africa (Kenya, Uganda and Tanzania)

be based on the type of energy required for immediate use: heat or electricity.

Let us first consider domestic energy requirement. The top priority is heat energy mainly for cooking. The second priority is lighting that, like cooking, can be provided from a number of sources: oil, electricity and biomass. All the three sources provide sufficient heat for most applications but only oil and electricity provide high quality lighting. The costs of oil and electricity are however higher than the cost of biomass which is normally freely collected in most parts of the rural areas. Thus a middle-income family living in an urban set-up would make a completely different energy choice during vacation in the rural home. The point is, the choice about which type of energy source to use is more than economic consideration. It is a very tricky and delicate mixture of economics, availability and application and this is not limited to domestic energy needs only but it is also relevant in industrial energy considerations. It is for this reason that a low income family living in the city would crave for electricity because of the desire to power household appliances which are the mode of city life but would use a different energy source for cooking and heating. On the other hand, a high-income city family may use electricity for cooking, lighting and powering household appliances but would use wood for space heating.

Industrial power considerations also look at a number of issues including industrial raw materials needed in the production line. For example, a sugar industry accumulates a lot of bagasse while at the same time it requires a lot of heat in its production boilers. The boilers produce steam some of which finally end up in the environment. In such a situation, the choice of fuel for the boilers would naturally be the abundant sugar cane bagasse and since steam is already one of the bye-products, the factory can go further and use the steam to generate electricity for its own use. On the other hand, if the industry needs only electrical power to run its machines, then it is more sensible to get the supply from existing power grid. A brick factory may need electricity for its support systems but use wood in the brick-firing kilns. It is therefore reasonable to conclude that energy choice is made on the basis of a number of factors and not just cost.

In general and in global terms, the level of industrial and economic development of a country determines the predominant source of energy and the range of other options available. Much of the energy used in the world today are exhaustible and are gradually becoming scarce while energy consumption rate is steadily increasing. More than 50% of the energy that has been consumed in the last 2000 years was consumed in the last 100 years. During this period, there has been a significant shift from one primary source to another particularly in countries that experienced and benefited from the industrial revolution. For example, in 1950, coal was the major global source of energy while oil was a distant second source as shown in Fig 1.3.

About 20 years later (1968), the contribution from petroleum had substantially increased while that of coal fell by a large margin (Fig. 1.4). Both petrol and natural gas were rapidly replacing coal as the source of energy for space heating, electricity generation, transport and cooking. New technologies, particularly nuclear, were developed with more attention given to the fact that petroleum and natural gas were set to be the major sources of energy for transport vehicles and other

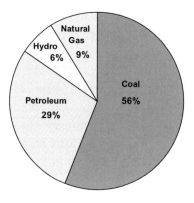

Fig. 1.3. Contributions of Commercial Energies in world energy supply (1950)

mechanical machines and, as a result, technologies such as steam engines that relied on coal began to decline. These changes took place more rapidly in the industrialized nations. For example, in the United States of America, use of petroleum increased from 14% in 1920 to 50% in 1980 while use of coal decreased from 78% to 18% in the same period. The emergence of nuclear power also stifled further development of large hydro power plants. By 1980 both petroleum and natural gas contributed more than 75% of the total energy consumption in the USA. These changes occurred principally due to increased technological and industrial developments that substantially improved the economies so that more and more people were able to afford and hence demand cleaner energies. The two events (industrial development and increased consumption of clean commercial energies) are mutually dependent so that one leads to the other and vice versa. The large consumption of coal in those days is comparable to the present high consumption of wood fuel in the developing countries particularly in sub-Sahara Africa (including South Africa) as shown in Fig. 1.5. If South Africa is excluded then the average contribution from biomass would be as high as 80% in Central and Eastern Africa region. Solid fuels include mainly coal from South Africa and some small quantities from the neighbouring states of Zimbabwe, Botswana and Mozambique. However, wide scale biomass use is not restricted to Africa. Its contribution is reasonably large even in global terms since more than 50% of the total world population still rely heavily on it. The use of biomass is largely limited to heat requirements, as it is unsuitable and inefficient when used for other purposes such as lighting. This naturally restricts its consumption and therefore, in global terms, the poor majority who depend on it consumes less than 20% of the total world energy consumption. To understand this, it is important to consider the total world energy consumption from all sources and compare this with typical energy consumption in a developing country. Energy consumption in an East African state of Tanzania is presented in this comparative analysis but first the general situation in sub-Saharan Africa is exposed as shown in Fig. 1.5.

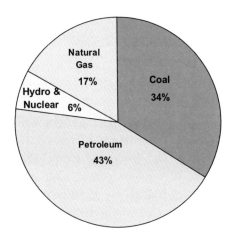

Fig. 1.4. Commercial Energy resources and their world contributions in 1968

Sub-Saharan Africa is a region of diverse energy resources that include oil deposits in western and central parts mainly in Nigeria, Angola and Southern Sudan; large deposits coal in southern Africa and natural gas in some countries. There are also geothermal energy resources particularly in Ethiopia and Kenya. In addition to these, some of the major African rivers like Nile, Niger, Volta, Congo, Zambezi, Limpopo and Orange with huge hydro potential are in the Sub-Saharan region. All these make very little contribution to the total energy consumed in the region compared to biomass energies as shown in Fig. 1.5. Much of the oil consumed is in fact imported from outside the region while hydropower development is still below 10% of the available potential.

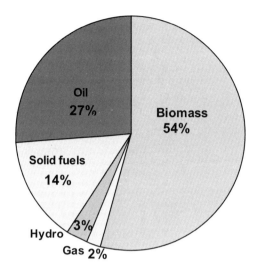

Fig. 1.5. Energy contribution by source in 1990 for Sub-Saharan Africa

Let us first look at the total world energy consumption by source in order to understand the role of biomass energy.

The information given in Fig. 1.6 is based on the total world energy consumption, which was about 9.9 billion tons of oil equivalent in the year 2000. It is clear that biomass is still a significant source of energy even when all modern global energy sources are considered. Geothermal sources contributed about 0.46% while the others included in that quarter (wind, solar and tidal) contributed only 0.04%. The situation in East Africa is remarkably different from that of the rest of the world and to illustrate this, we look at the Tanzanian case, which represents moderate situation in East Africa. In the same year (2000), Tanzania's total energy consumption was only about 15 million TOE distributed as shown in Fig. 1.7. As has been severally mentioned, the rural population of the developing countries can be as high as 95% of the total national population and almost all of them depend on biomass energy. Thus, apart from commercial energies used in transport and industrial sectors, biomass remains the main energy source. Its contribution however is generally higher than household consumption because it is also

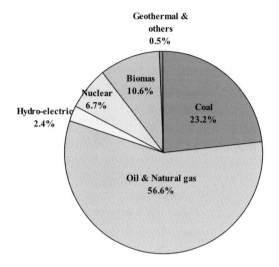

Fig. 1.6. World Energy Consumption by source in 2000

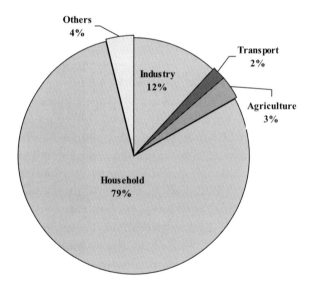

Fig. 1.7. Energy consumption by sector in Tanzania (2000)

used in industries, agriculture and for other non-household applications. This is clearly shown in Fig. 1.8 where national biomass contribution for all sectors is 93.6%. The other sources of energy on which the government invests a substantial amount of money, contribute much less than expected and this is also indicative of the very low level of industrial establishments and poor transport system. Fossil fuels include oil products, coal and natural gas. In East Africa, Tanzania is the only country with some deposits of natural gas and coal. Both Uganda and Kenya have neither coal nor natural gas but since these have not made any impact in Tanzania, the energy consumption pattern is comparable to those in Uganda while Kenya, with more developed industrial and agricultural sectors, spends relatively more energy in these sectors. For example, Kenya's total energy consumption in the same year was also about 15 million TOE out of which biomass contributed about 79%. Household energy consumption was however down to 68% while transport, industry and agriculture consumed 12%, 11% and 7% respectively. Hydroelectric contribution was only about 1.8% while imported fossil fuels accounted for about 19%. So the pattern is basically similar with only small variations in sectoral consumptions.

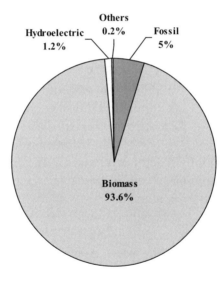

Fig. 1.8. Energy contribution by source in Tanzania (2000)

The high dependence on biomass energy is a clear indication that the governments have not paid as much attention to energy development as they should. This may be one of the reasons for the very slow rate development process in the region. It should cause some concern not only for the governments but also for environmentalists since continued use of biomass has adverse consequences on soil fertility and environment in general.

The big question for East Africa is whether the region whose average wood fuel consumption stands at about 85% of the total energy can switch to cleaner energies, as did the USA and Western Europe. Since East Africa is not endowed with oil or sufficient natural gas, what would be the feasible alternative sources considering the fact that petroleum and natural gas can only be supplied from capital reserves? At present, the future of the economy that would generate that capital reserve is not clear for the East African case.

1.4 Household Energy Application and Management

Energy can be discussed from two viewpoints: its application and its source. For its use, there are priorities which people tend to follow naturally particularly in household application where energy for cooking takes top priority followed by energy for lighting. Applications beyond these depend on family income. Whereas commercial energies are well managed, household energy demand and use is more complicated and are often affected by factors beyond government control. In East Africa, household energy accounts for over 80% of the total energy requirement and so it is important to understand energy applications and sources for this sub-sector. East African countries are in a region of rich diversity in terms of geography, peoples, income levels and natural resources. Depending on family income level and traditional eating habits, household may prepare and eat up to four meals per day. This includes tea, porridge and other light meals. Some foods can take as long as four hours to cook while others may be prepared in a shorter time. About 95% of the rural population rely entirely on wood for cooking using the traditional three-stone fireplace. A few households use the improved version of the three-stone in which the fire area is insulated with mud to reduce heat losses. A small number of households use more efficient baked clay stoves that were introduced and disseminated mainly by NGOs during the last twenty years. Wood is the main fuel for these stoves although the use of cow dung and crop residues is practiced in areas where wood is either scarce. Charcoal is the next important fuel for cooking mostly used by the urban poor but there has been a steady rise in the use of charcoal in the rural areas, especially areas where wood supply is well below demand. Charcoal is burned in simple metal stoves or improved ceramic-lined stoves with better fuel consumption efficiency. Local artisans make both stoves. Kerosene is widely available but is used only occasionally for quick cooking purposes because it is more expensive than biomass-based fuels and requires expensive wick stoves. Even those who can afford it hardly use it because of safety and the fact that food cooked on kerosene stove sometimes absorbs the fume that gives irritating smell to the food. The use of LPG is very limited particularly in Uganda due to the high costs and inadequate supply in the rural areas. Electricity similarly plays a very minor role as household energy for cooking principally due to its high cost and limited accessibility. Both Uganda and Kenya do not have coal reserves and therefore it is not used as household energy but some quantity is imported for industrial applications. There are some coal mines in southern Tanzania but it has not become a common household energy due to the general lack of marketing and infrastructure for it. Thus, biomass (wood, charcoal, crop residue and cow dung) is clearly the most important source of household energy for cooking in East Africa.

Lighting is the second most important priority area for which energy is required. Although the available energy sources for this are expensive, most rural households can still afford to use some of them because of the short duration of lighting requirement. Each day, most families need light for an average of about three hours. Furthermore, cooking with firewood provide some light around the cooking area so that, sometimes, additional lighting may not be required if the whole family sits around the fire. On moonlight nights, cooking can be done outside the house and so reducing the need for lighting. Most households use kerosene as the source of lighting energy, using simple wick lamps with or without glass windshields. Kerosene consumption of these lamps is very low, making them very attractive to most rural households. Over 95% of the rural population use these lamps even though kerosene itself is expensive for them. Use of the more expensive dry cells is also common but their application is limited to flashing only, when it is desirable to identify an object. Some lead acid batteries are used for lighting and entertainment but this is severely restricted by both the cost and lack of recharging facilities within easy reach. Similarly, solar photovoltaic systems have not made any significant impact as rural lighting energy for a number of reasons including high initial cost. All other household energy needs such as for radio, space and water heating are constrained

by poverty and are therefore mostly unmet for the majority of the low
income households.

Energy sources will be discussed in Chap. 2 but a brief mention
now will suffice to shed some light into the underlying problems in the
supply of household energy. The region has a wide diversity of energy
sources, which are used for similarly diverse applications. The use of
the various sources is prominently constrained by poverty. On average,
about 45% of the population live on less than one US Dollar a day but
about 90% live on less than three US Dollars a day. This means that
most people cannot fully meet their basic needs such as decent meals,
clothing, health care, energy and sanitary residential conditions. They
have to struggle to obtain most of these by employing own production strategies and heavy reliance on natural products especially for
energy and health care (biomass and herbal medicine). This is why
wood, supplemented by cow dung and crop residues, is overwhelmingly
the single most important energy for cooking in the rural areas while
charcoal is the dominant source for the urban poor. Most rural households (about 90%) collect wood from the surrounding areas without
having to pay for it while the others may mix buying and collecting.
Urban households, on the other hand, buy all their charcoal and some
small quantities of wood both of which are transported from the rural
areas into the towns where there are several outlets, mainly within residential estates. The prices vary but usually remain affordable for the
low-income groups. Crop residues and animal dung are exclusively for
rural application where they are generated and used within the family's
land and homestead. They are obtained freely and are not considered
as commodities for sale. In some tribal customs, it is even a taboo to
sell cow dung and agricultural waste and this means that anybody can
collect them from anywhere.

Table 1.3. Annual consumption of basic household energies in million TOE

	Wood	Charcoal	Kerosene
Kenya	4.1	0.6	0.3
Uganda	11	1.5	0.02
Tanzania	5.3	2.3	0.04

Table 1.3 shows the common sources of household energies and their
relative contribution in TOE. Although these consumptions have some
bearing on differences of national populations, it is evidently clear that

Kenya consumes far less wood and charcoal than Uganda and Tanzania but its consumption of kerosene is much higher than in the two countries. One possible reason for this is the availability of oil refinery in Kenya and Kenya government's low tax on kerosene in addition to better distribution network in the rural areas. Kenya's per capita electricity consumption is also high, indicating that Kenya is already ahead of the other East African countries in the use of commercial energies.

Whereas commercial energy management is closely monitored at the national levels, there appears to be no overall or even coordinating policy at household energy level. Even the statistics on household energy consumption is not detailed and is usually collected by the national bureau of Statistics as an appendix to the national census records that are collected once in a decade. In addition, information on energy mix in terms of ratios, problems and constraints are usually not collected. Of course, the national policies of all the three East African states do consider household energy demand and supply but do so only as a responsibility and not as a commitment. The emphasis and action are on commercial energies and some concern is on sustainable use of biomass energies but virtually nothing is done to make clean household energy affordable. If energy is unaffordable, then it is not available. Uganda is the only country in East Africa that is concerned about affordability of energy for the rural households and has introduced measures to make energy available to rural households. Some of these include use of 'smart subsidies', prepayment meters, use of load limiters and leasing of PV systems. Tanzanian consumers are already using prepaid meters while all these are still alien to the Kenyan consumers. In general, the confusion with which rural household energy is managed arises from national resource management arrangements, which tend to scatter household energy issues in different government ministries. For example in Kenya, biomass issues are handled by the ministries of Environment, Agriculture and, to some extent, ministry of Energy, and each of these ministries may not be really concerned about energy par se but other functions of biomass as well.

1.5 Commercial Energy Management

Oil fuels and electricity are the two main energy resources, which are of both commercial and political interest in East Africa. Oil includes diesel, petrol, kerosene, aviation fuel and LPG and is a versatile energy source that can be used to produce most of other forms of energy including the generation of electricity. Tanzania has some coal with significant commercial value but its contribution in national energy scene, at present, is very limited. It will not therefore be discussed in details. Up to the year 2005, there were no known exploitable oil reserves in the region but active exploration has been, and is still going on within the three East African states of Kenya, Uganda and Tanzania. These are going on under various agreements between multinational oil companies and the individual state governments. East Africa is the only region in Africa that is a net crude oil importer and Kenya is the region's largest consumer and ranks overall 9th in Africa. The region therefore imports all its oil requirements mainly from United Arab Emirates in the Middle East. In 1988, an oil company found oil at a depth of about 4000 meters in Isiolo district in Kenya but it was considered economically unviable to exploit. Energy use for commercial and industrial purposes is well managed through an organized demand-supply system controlled by both governments and the private sector as the major stakeholders. Prior to the so-called energy sector liberalization of the late 1990s, oil was imported into the region in its crude form and then refined at the coastal city of Mombasa in Kenya and also in Tanzania. The importation of oil was closely controlled by national oil companies which were allocated both import and distribution quarters. The Tanzanian refinery was later closed down and converted into storage facility on the basis of expert recommendation. Uganda is landlocked, has no oil refinery facility and has to depend on imports of refined products mainly through Kenya whose refinery is capable of processing about 90,000 barrels of oil compared to the Tanzanian refinery, which had a capacity for only 14,900 barrels per day. Both these refineries had not only been operating below capacity but had also been faced with management and financial problems. Obsolescence is also an issue for the Kenyan refinery for which up-grading overhaul has been recommended. The Kenyan oil refinery was farther affected by the liberalization of the energy sector that allowed oil companies to import refined oil directly from the oil producing countries. As the three East African states get closer together again through the re-established East African Community, several agreements have been put in place to improve the energy sector. For example, Uganda and Kenya have agreed to extend the Kenyan oil pipeline from the town of Eldoret in western Kenya to Kampala, Uganda (320 km) in order to efficiently serve Uganda and its neighbours such as Rwanda, Burundi, eastern parts of the Democratic Republic of Congo and northern part of Tanzania. Tanzania is also to build an oil pipeline from the depot in Dar-es-Salaam to Indeni refinery in Ndola, Zambia, a distance of 1,710 km. The pipeline that has a capacity of 1.1 million tones annually is to be owned jointly by governments of Tanzania (33%) and Zambia (67%). There is also a plan to construct another pipeline from Dar-es-Salaam to the Lake Victoria city of Mwanza in northern Tanzania (1,104 km) to serve Uganda, Rwanda, Burundi and Democratic Republic of Congo. Since oil consumption is steadily increasing in the region, Kenyan refinery at its present capacity will not be able to adequately serve the region. Even if it is upgraded to a higher capacity, it will still not be sufficient for the future increased demand. East Africa presently consumes an estimated 90,000 barrels per day with Kenya as the major consumer at about 60,000 barrels per day. Although Uganda consumes less than one quarter of Kenya's consumption, it spends about 12% of its total import bill on oil importation while Kenya's consumption is about 20% of its import bill and so, in general, importation of oil by the three East African states is putting a lot of strain on their economies. Table 1.4 shows some oil related statistics in the three countries - consumption, percentage contribution to energy and percentage of import bill.

Table 1.4. Some oil-related statistics in East Africa

	Oil consumption (million barrels per year)	Percentage of total energy	Percentage import bill
Kenya	20	26	20
Uganda	3	7	12
Tanzania	5.5	7	n.a

Since 1980, the prices of oil products and commercial energies in general have significantly gone up, forcing the prices of other commodities to also increase. This has created socio-economic problems in the region as more middle-income groups fall back below the poverty line. The price of kerosene, used mostly by the low-income groups, has increased by about 170% while charcoal has gone up by 60% since 1980, pushing low-income groups farther into the fathoms of poverty. The costs of LPG and electricity have also increased by 75% and 70% respectively. These increases are calculated against the prices of the fuels in US dollars and the value of the dollar is assumed to have remained constant during the same period. Fluctuations of oil prices are the most dynamic among all commercial energies as both international market forces and internal politics affect them. This is an area that must be considered carefully as it can be the source of a serious internal oil crisis. For example, in the late 1990s oil prices sharply went up and the governments blamed this on the market collusion by the multinational oil companies to increase their profit margin. These companies, on the other hand, considered this as an attempt by the governments to divert public attention from their poor economic management to the energy sector. This particular conflict led to the establishments of regulatory boards by Kenyan and Tanzanian governments to oversee the oil industries. In order to assess the effectiveness of such regulatory bodies, we look at a case in 2005 when the Kenya government reorganized the method of paying taxes on oil imports while keeping the tax level unchanged. The new tax payment method required oil companies to pay the tax upfront The oil companies did not support this method and immediately raised the pump prices so that motorists had to pay more even though the local currency had made significant gains against the US dollar, lowering the import bill for the oil companies. This was a clear demonstration of how the cartel of oil companies can use their strength to maximize their profits against the government's regulatory mechanisms. This happened because the limited storage capacity is not capable of holding enough oil for a long time and therefore the government cannot afford to push oil companies to disrupt import supplies. The governments can be blackmailed by the oil cartels and so the regulatory mechanisms in this sub-sector should be reviewed to introduce an arrangement that would make sure the government is not held at ransom during difficult times. Another interesting observation, which brings to doubt the pricing factors such as transport costs, is the fact that the pump price near the refinery facility is sometimes higher than the pump prices 800 km away. So although the governments are very sensitive to any events that are likely to affect the supply of oil into the region or that could adversely affect oil prices, there is very little they can do to locally control the situation. It is for these reasons that internal fluctuations of oil prices often cause sharp conflicts between multinational oil companies, which import and distribute oil in the region and the governments. Thus the looming global oil crisis that is likely to continue beyond 2005 should cause serious concerns in East Africa and some early measures may be useful. However, despite the occasional conflicts that revolve around protection of interests, the governments and the oil companies have worked very closely on acquisition and distribution of petroleum products. Kenya Petroleum Refineries Limited (KPRL) is 50% owned by the government of Kenya with remaining stake held by oil companies (Shell and BP having 17.1% each while Chevron holds 15.8%). In 2005, the government announced its intention to spend up to USD 200 million to upgrade the refinery, which has been using hydro-skimming technology that produces large quantities of low-value fuel oil and insufficient high-value products. The proposed rehabilitation will enable the plant to produce unleaded fuels. Obviously, some of this expenditure will be passed on to the consumers through appropriate pricing of petroleum fuels in the local market - a move that will further intensify economic hardships that are already caused by the high prices due to increased worldwide demand for oil. In order to understand the management of petroleum as an important energy resource in East Africa, we take an example of how it has evolved in Kenya.

The Kenyan domestic consumption of petroleum products stands at an average of about 2.6 million tones annually and all of this is imported. The point of quantitative control is at the Kenya Petroleum Refinery Limited (KPRL) at the seaport of Mombasa located in the south east of Kenya on the shore of Indian Ocean. The refinery was incorporated in 1960 under Shell and BP oil companies but effectively started operating in 1963. Later more oil companies joined by buying shares in the company and in 1971 the government of Kenya acquired 50% equity. The refinery uses technology that produces large quantities of low-value fuel oil and relatively less high-value white products. From the refinery, the products were initially transported to Nairobi, about 500 km away, through 35.55 cm (14-inch) diameter pipe but this was later extended to Kisumu and Eldoret in western parts of Kenya using smaller pipes. The pipelines transport about 60% of all the petroleum products while Kenya Railways and several road transporters share the rest. The distribution is done through countrywide network of licensed petroleum stations owned by registered oil companies. Some of the products are exported to the neighboring countries. Before the so-called liberalization of the sub-sector, the schedule of prices of petroleum products was developed by the minister for Finance in consultations with the Ministry of Energy and circulated to the oil companies for implementation. A number of factors such as procurement cost, processing, taxes, transport and profit margins were considered in working out the prices. However the pricing system was considered to be opaque and did not appear to have been based on sound economic considerations and therefore was considered to be a government scheme to manipulate demand in order to cope with imbalances. This created considerable suspicion and animosity between the government and oil dealers. There was need to develop longer term pricing structure that would enable oil companies to get adequate returns for the investment and plan ahead for expansion. These were some of the issues that forced the oil companies to demand for liberalization of the industry so that the supply and demand forces are allowed to determine the market prices. When this was finally done in the mid 1990s, oil companies were required to ensure that there is enough oil stock in the country to last thirty days and LPG for at least ten days. They were also allowed to choose to either import processed oil or use the refinery in the country in addition to other conditions. All aspects of government monopoly were removed and this encouraged new investors into the oil industry. Although the old companies still control about 75% of the domestic distribution and retail markets, there are at least ten new entrants who are active in importation and distribution of petroleum products. The liberalization, however, seems to have increased insecurity of supply and weakened government control of the industry.

The management of the electricity sub-sector is somewhat different from that of oil. Before the three East African states got their independence, the region was administered by the British government and the main source of electricity for both Uganda and Kenya was the hydro power station at Jinja in Uganda. One of the agreements made at the time of independence was that Uganda would continue to supply at least some 5% of electricity to Kenya even if Kenya finally developed self-sufficiency in electricity production. To date, Kenya still gets a small quantity of electricity from Jinja although there are indications that Uganda is likely to terminate this arrangement due to increased electricity demand at home. Apart from this, all the three states generate their own electricity principally from hydro stations owned and managed by the state. In addition to hydropower, Kenya gets some of its electricity from geothermal facility located in the Rift Valley province. Other sources of electricity include large portable generators some of which were installed by independent power producers that were given licenses as a result of the liberalization of the power sector. Arrangements have also been made with Emergency Power Producers (EPPs) who specialize in supplementing power during crisis as a result of unexpected long period of drought. They normally use fossil fuel powered generators. Many of the current independent power producers emerged during the severe drought of 1997 when Kenya hurriedly invited EPPs to supplement the dwindling local production. All EPPs left after about four years of operation in the country. It is important to mention here that liberalization of the power sector was not the initiative of the governments but was part of the Structural Adjustment Programme proposed and supported by principal international development agencies notably World Bank and IMF.

Almost all power producers inject their contributions into the national grid for distribution by a partially state-owned power company. In Kenya it is the Kenya Power and Lighting Company (KPLC) that managed the sale and distribution of electricity while in Tanzania and Uganda, Tanzanian Electricity Supply Company (TANESCO) and Uganda Electricity Board (UEB) respectively handled both generation and distribution. These companies or para-statal organizations, as they are commonly known, have in the past faced severe management and financial problems due to government interference particularly with regard to sourcing of funds from the electricity company for non-energy related developments and haphazard political decisions that adversely affect the management of the power sector. Some of these included not just over-employment but also engagement of unqualified people in the sector. Such problems are often spurred by politically motivated favours to specific population groups. Monopoly status of the said power companies has also had negative impact on the development of the independent and alternative power production.

Proposal under the liberalization programme suggested that the power companies should be split so that generation and distribution are managed by different companies and, at the same time, allow independent producers to have certain rights under various agreements. Such proposals may bring some improvements in the electricity subsector but are unlikely to solve the management problems. Nevertheless, some steps have been taken to implement the proposal and, in this respect, Kenya Electricity Generating Company (KenGen Co) has been created out of KPLC to handle generation of electricity in Kenya. Uganda too has made some changes that created other companies to share the responsibilities of UEB with Uganda Electricity Transmission Company (UETC) dealing with transmission while Uganda Electricity Generation Concession (UEGC) handling generation. Similar trend is taking place in Tanzania where TANESCO is split into three different companies each to oversee generation, transmission and distributions respectively. However, unlike the oil sub-sector, the governments are still expected to have significant control of the electricity sub-sector. It should be remembered that the establishments of the existing power generating facilities were made possible through loans or arrangements for which the governments are still responsible. Therefore whatever adjustments or restructuring taking place, the governments must still have significant roles to play in running the services.

In addition to the boards that oversee the management of specific energy sub-sectors, all commercial energies fall under specified government ministry. Due to frequent changes in which the number of government ministries are either reduced or increased, commercial energies have moved from one ministry to another or merged with other ministries over the years. One clear thing, however, is that all the three states have recognized the importance of these two sources of energy and have made sure that they are handled, at the top, by a government minister. Although the companies that run the two energy sources have the authority to manage their affairs in their own best interest, the governments expect them to also provide the services and goods to the people efficiently and at 'reasonable' prices. It is this relationship that has made it difficult for some of the companies particularly the electricity companies to meet both their economic and management desires. Structural Adjustments and liberalization principles require governments to deal with provision of enabling environment and leave private sector to handle provision of services. Energy sector, however, is so sensitive and important that the governments are not willing to give away all their controls over the sector. This dilemma is partly the reason why this sector has not had the expected growth rate in the region. Potential investors have all along looked at government involvement with a lot of suspicion. Historically, the three companies that have been in charge of electricity generation and distribution in East Africa have their roots in the East African Power and Lighting Company (EAP & LC), which was the sole company in charge of generation and distribution of electricity in East Africa. This company existed before the then East African Community broke up in the late

1970s and it had a vertically integrated monopoly in the generation, transmission and distribution of electricity in the region. With the disintegration of the Community, each state constituted its own power company, which basically adopted the management structure and the monopoly policy of the defunct EAP & LC. Because of this historical background, the regulatory mechanisms in the electricity sub-sector in the three states have remained largely similar. In all cases regulatory authorities are established by Acts of Parliament and therefore provide strong government presence in the electricity production and management. The policy deliberately discouraged any independent operator from investing in energy production, as it would be impossible to sell it. In developed economies such as USA and Europe, the policies are quite different and anyone is allowed to generate own power and sell to the existing grid as long as the method used is safe and environmentally friendly. This demonstrates the difference a good policy can make if effectively implemented.

1.6 Energy Choice Factors

As developing countries, Kenya, Uganda and Tanzania still have to do a lot in order to provide basic requirements to their people. The 2005 estimates put the region's population at about 90 million people, 80% of whom live in the rural areas. There is however a number of small but heavily populated settlement areas commonly known as Market or trading centers which are expected to eventually develop into Townships. Most of them spring up along major road networks due to trade prospects with travelers. They are nevertheless considered as part of the rural set-ups since they do not have organized service provision systems such as water, sanitation and infrastructure. Furthermore the people's life style in these places is basically rural in nature.

Generally, national grid power lines tend to follow major road networks for ease of maintenance and security and therefore many so-called market centers have access to electricity but a good number of their residents are not supplied with electricity despite its accessibility. This immediately raises questions about the factors that consumers have to consider in making energy choices. The generally poor infrastructure implies poor distribution of electricity. The improvement of infrastructure has proved to be too expensive for the governments due to the fact that rural settlements are so sparsely scattered that any communication system would require very complicated and expensive network. Population density can be as low as 20 people per square kilometer and even much less in some places. This low population density, coupled with the difficult terrains makes it extremely expensive to extend the electricity network to most parts of the rural areas especially if generation of power is centralized as is the case in East Africa. The distribution of oil-based fuels (petrol, diesel, kerosene and LPG) is also severely constrained by poor infrastructure. However, its transportability in the refined form enables traders at various levels to get involved in its commercial distribution, increasing its cost at every stage but giving the people the much needed opportunity to obtain it a little at a time. It is therefore too expensive for the rural communities whose monthly incomes are not only low but also is regular in many cases. The concern in energy supply therefore goes beyond the availability of a preferred source of energy and encompasses the preparedness and ability of the rural communities to benefit from such energies. This could be done by appropriately packaging energy to suit both energy requirements and economic circumstances of the rural communities. Availability of energy when people cannot afford it does not make any sense.

Traditionally, energy supply for domestic applications among rural communities is the responsibility of women who can only recruit children to help. This has been so because the main source of energy has always been biomass materialssuch as wood, cow dung and agricultural wastes which are largely obtained free of any charge. These were considered to be readily available and only needed to be collected and taken home. However, it is the simple tasks of collecting firewood, fetching water, preparing food for the family, taking care of children, cleaning the house, cloths etc that have over-burdened the women to the extent that they could not fully participate in many socio-economic developments such as education. As a result of this heavy responsibility of managing practically all household chores, rural women have remained at the bottom of the social strata with virtually nothing for their welfare or under their ownership and time spent on collection of energy resources has been one of the key factors in all this.

Despite the economic development and gender equality policies, the energy situation among rural communities has not changed and has continued to be a burden on women and children. Biomass is still the major source of energy for both cooking and lighting for most people in East Africa. Rural electrification coverage in the region is on average less than 2%. If both urban and rural population are considered then a total of 10% of Kenya's population have access to electricity while only 6% have access in Uganda. The situation is not likely to change in the foreseeable future and, with the dwindling biomass resources, the region is likely to face a severe energy crisis.

Several decades ago, abundant natural forests freely provided the wood fuel for practically all domestic energy requirements. As population increased, the demand for more land for settlement and food production also increased and naturally forest cover diminished. Women began to travel longer distances to fetch wood fuel and this made it necessary to start growing trees specifically to provide domestic energy. Soon there was a general concern for wood fuel supply and its conservation. Consequently, there was an increase in extension services on agro-forestry with significant support from various development agencies. Such programmes considered all aspects of wood requirements including improvements of efficiencies of wood and charcoal stoves. As a result of these activities, new designs and presumably more efficient stoves were introduced into the market. Despite all these efforts, the role of wood as a major source of rural energy has not changed but the use of charcoal stoves among rural communities has increased largely due to the success of the so-called improved ceramic stoves, which were first developed in Kenya. In general, however, energy distribution patterns by source are basically similar in the three states of East Africa and are likely to remain so unless drastic and perhaps revolutionary measures are adopted in the energy sector. Biomass is expected to remain the major source of energy in East Africa (Table 1.5). The high per capita electricity consumption in Kenya is attributed to the relatively high level of industrial development and the extent of commercial energy distribution particularly kerosene. Other sources of energy such as wind and photovoltaic solar systems, which can provide electricity to the rural communities, have made relatively small contributions despite the abundance of both wind and sunshine.

Although biomass is often associated with rural household energy, a significant amount is also used in agro-based industries such as sugar, tea, tobacco and brick-making industries. Rural schools and hospitals also consume a lot of biomass energy for cooking and heating.

Kerosene is considered to be too expensive for the rural poor. However, it is used by many rural households almost exclusively for lighting, for less than three hours each day. Therefore monthly expenditure on

Table 1.5. East African Energy Contribution statistics in percentages

Source/Type	Kenya	Uganda	Tanzania
Electricity	3	1	1.5
Oil	26	7	7
Biomass	70	92	91
Others	< 1	-	0.5(coal)
National household electrification levels	13	6	8
Per capita Electricity consumption/kWh per annum	125	50	60

kerosene is kept as low as possible. This situation masks an unmet need, which is demonstrated when higher quality lighting is required. For example, when rural homes acquire biogas or electricity, lighting hours increase and so do evening activities that can improve their welfare. The use of kerosene by poor rural households despite its cost, strengthens the case for appropriate packaging of energy that enables consumers to get it in small quantities at a time. Rural application of LPG is limited and is restricted to the few high-income families who can afford it and have the means to fetch it from sparsely distributed rural petrol stations.

East Africa is a region of diversity in terms of geography, cultures, ethnic groups and level of income. The lifestyle of the people and the preferential use of energy resources are largely determined by family income but this is not the only factor. In general the household energy demand is for two basic requirements: cooking and lighting. When these have been adequately met, then the family can, if affordable, demand more energy for household appliances such as radios, torches and small television. Table 1.6 shows percentage energy contributions from various sources and presents an example of their prevailing uses by rural and urban households. The example is taken from Kenya where, on average, biomass energy contribution is about 80% with petroleum and electricity contributions at 18% and 1.4% respectively. Considering all sources of energy, rural households consume 57% while urban households and commercial sector respectively consume 16% and 27% of the total national energy consumption. The situation in Uganda and Tanzania follow a similar pattern and therefore these figures are good indications of the situation in the region.

Table 1.6. Total energy share in percentages for all sectors (Year 2000)

	Firewood 36.3%	Charcoal 38.1%	Industrial wood 0.3%	Wood wastes 0.5%	Farm residue 5.3%	Elect-ricity 1.4%	Petro-leum 18%	Total 100%
Rural Household	89	46	62	100	5	3	57	
Urban Households	2	36		38	26	5	16	
Others (Transport, commerce, Agriculture)	9	18	100			69	92	27
Total in %	100	100	100	100	100	100	100	100
Totals in '000'Giga Joules	252,000	264,000	1,750	3,520	37,000	9,800	125,000	693,070

The choice of which source to go for would depend on affordability and availability. Energy for cooking is the first priority and therefore it must be available all the time. Light is needed but if there is no energy for its provision, a family may do without it for a few days. However firewood does provide some amount of space lighting around the fireplace so that additional light source may not be required. Often, one source would be for cooking and another for lighting. For example in urban areas, the combination may be LPG (cooking) and electricity (lighting) or electricity (cooking and lighting) for high-income group while the middle-income group may choose charcoal and kerosene (cooking) and electricity (lighting). The urban poor would go for charcoal (cooking) and kerosene (lighting). Some urban elite would use wood for space heating but this is not based on economic considerations. In the rural set-up, availability and security of supply is an important factor and therefore the majority would use a combination of wood and kerosene for cooking and lighting respectively. Kerosene is preferred for lighting and is used only for short periods of 2 to 3 hours each night and so its consumption is low and hence low expenditure on it. Attempts have been made to introduce a variety of sizes of LPG storage and application devices in an effort to increase affordability but this has been constrained by the general high level of rural poverty and doubts on its immediate accessibility when it runs out. Energy is so crucial that people want to make choices that will ensure that it is always there when

needed. Table 1.7 shows total energy consumption by all sectors in East
Africa and also indicates preferred fuel mix for various applications

Table 1.7. Preferred fuel mix in percentages for three main consumption categories
in East Africa (2000)

Fuel use category	Fire wood	Char-coal	Industrial wood	Wood wastes	Farm residue	Elect-ricity	Petro-leum	Total
Rural Household	56	32	-	0.5	10	1	0.5	100
Urban Households	5	90	-	1	0.1	2	1.9	100
All other sectors	12	27	0.5	1.5	20	36	3	100

'All other sectors' include users that cannot be categorized as household such as cottage industries (brick-making, jaggaries, fish smoking,
small bakeries, milk processing etc), transport, agriculture, commerce
and industry. Charcoal consumption outside households is generally
found in cottage industries such as commercial food kiosks and rural restaurants while schools, hospitals, tobacco and tea processing establishments use firewood. Farm residues include sugarcane bagasse
used by sugar milling industries for steam production mostly in co-
generation systems. The percentages are calculated on the basis of energy consumed in joules from the chosen source. For example, rural
household firewood consumption is estimated to be about 700 million
Giga Joules while petroleum and electricity consumption by all other
sectors are 190 million and 15 million Giga Joules respectively.

The three main sources and types of energy (electricity, oil and
biomass) that are commonly used in East Africa are acquired through
certain technologies that determine the extent to which they can be
used in different places in the region. The choice would be based partly
on the economic activities and the required final form of energy. The
two forms that are commonly in demand are heat and electricity. Therefore the popularity or wide use of a particular form is determined by its
convertibility to other forms and the required applications. For example, electricity is used for powering industrial machines and household
appliances and can also be readily converted into heat and light. Similarly, oil-based fuels can be readily converted into heat and light but are
not suitable for powering household appliances. Solid biomass, on the
other hand can be converted into heat and light, but its quality of light

is very low and rapidly consumes biomass materials. Of course there are possibilities of obtaining some form of oil and gas from biomass but these technologies are presently not widely used. Given these qualities, electricity would be the best choice because of its versatility followed by oil and then biomass. If the price is the determinant factor and the final application is heat-based then biomass would be the obvious choice followed by oil and then electricity. This is based on the assumption that the pricing includes only financial outlay for the fuel and its utilization appliances, and no consideration is given to non-monetary costs to the environment and the user. Finally, if the choice were based on availability and life style then the rural community would choose biomass, oil and electricity in that order, while urban population would go for electricity, oil and biomass. We see that in all these considerations, oil is the second energy choice while electricity and biomass switch positions depending on the local circumstances. These are the simple characteristics that energy planners often ignore and yet, for the end user they determine the choice of the energy source. Therefore, assuming that the economy of East Africa remains nearly stagnant as is often the case, the situations presented in Tables 1.5, 1.6 and 1.7 will prevail for many years to come. This is because people will always make choices that are appropriate to their economic, social and geographical circumstances. Another factor that is often ignored is the role of dry cells and rechargeable lead acid batteries as sources of electric energy for rural population. Individual ownership and hence effective application controls make these devices very attractive in the rural set-up. Thus, we see that with dry cells, rechargeable batteries, a little kerosene and biomass, a rural household is not only able to meet all its energy requirements at an affordable rate but is also able to control the sourcing and application of the energy. Actually dry cells, lead-acid batteries and kerosene, although used by many poor people, are more expensive than grid electricity for lighting. It is therefore important to note that poor people often unwittingly end up using energy that is too expensive , simply because it can be acquired in small amounts and their use can be controlled. Dry cells, for example, are used only as flash lights and so a set can last as long as one month or more. Since biomass is considered the only energy source that can be obtained free of any charge, it also serves an energy security. However, as discussed earlier, biomass resource is under threat, in spite of valiant efforts to address security of supply and improve efficiency of use. Hidden costs of biomass include the longer hours and increased drudgery for women in procuring biomass fuel and using it in rudimentary appliances; and adverse health impacts from smoky kitchens which are the norm in the rural areas.

1.7 Institutional Constraints

We have seen that industries and households whether poor or rich require either heat or electrical energy or both for various applications. Electrical energy is more preferred because of its conversion versatility and the fact that it is a high quality power source for industrial machines. The other sources cannot serve the same purpose or, if they do, would be too expensive. Oil however will continue to play its crucial role in the transport sector. Investments on commercial energies have been based on national funds, extended loans and grants but such resources are continually becoming scarce due to a number of reasons and this trend has increasingly become an inhibiting factor for energy development in the region. Such constraints are often used to persuade national governments to accept externally initiated policy changes in order to continue receiving development loans or grants from international development partners. The objective of such new policies is to offload management responsibilities from the government to the private sector to allow the governments to concentrate on creation of conducive business environment and not direct service provision to the people. At the same time such policies make the government's energy management more vulnerable to private sector influence, some of which may not be in the best interest of the state. The implementation of the liberalization policies in the energy sector in East Africa is a result of such policy changes. This is unfortunately happening at a time when most rural communities are still relying on wood as a major source of energy. Over reliance on large centralized hydro generating facilities has not provided the solution to this problem and very little effort, if any, has been made to implement viable alternative energy technologies. Consequently the cost of generation and distribution of electricity has not reduced at a sufficient rate to satisfy the economic conditions of the people. It is estimated that the average cost of grid extension in conditions of low population density and difficult terrains can be as high as US dollars 10,000 per kilometer. Under such circumstances, grid extension is not cost effective and alternative electricity supply systems need to be considered if viable grid extension policies cannot be found. There are variations in power extension policies in East Africa but, in general, the consumer is expected to meet the cost of grid extension. In addition to this there are other payments such as minimum charge that the consumer has to pay every month whether power is consumed or not. The rural poor cannot afford all these costs and therefore attempts to implement them will not promote rural electrification. The policy therefore sends a very wrong signal even to those who can afford these charges and unless it is reviewed in favour of the consumer, then even if the available hydro potential in the region were to be developed, still the majority of people living in East Africa would not be able to use electricity because it would be too expensive for them and so reliance on wood fuel would continue. Thus continuing with the present trend of using expensive centralized large hydro stations is not a viable alternative to biomass. In addition, large hydro stations create ecological conflicts and displacement of a large number of people in the area. They also have problems of silt, which fill the dams during the rainy season and prevent full capacity generation. Operation below capacity is also a known problem during prolonged periods of drought. Such problems will always exist in addition to distribution problems, which cause frequent power disruptions. In Kenya alone, about 10,000 cases of such disruptions are reported each month. Other methods and more user-friendly policies in electricity generation and distribution should be explored with more vigour and deliberate planning. Decentralized systems that give the user some responsibilities in maintenance and control of applications would be cheaper and more likely to appeal to the rural population. Oil, the other major commercial energy source, is also faced with similar problems of affordability but since the consumer can get it in small and hence affordable quantities and control its consumption, more and more people will be able to use it. This is a good example of how poor people end up using more expensive commodity simply because it is possible to obtain small quantities. However, increased demand for oil worldwide and the fears that oil reserves are dwindling will probably force the price to escalate beyond the reach of poor rural communities. The apparent increase in demand for kerosene as the main source of light among rural population may also force the price to go up and reduce its wide scale application. One would therefore wonder whether it is a good policy to promote the use of kerosene in a large scale or not. Increased demand would most probably trigger price increases that could severely hit the rural poor. It is a well-known fact that about 50% of East Africans are living below poverty line and they would not be able to afford the cost of either electricity or kerosene should the prices go up significantly. Thus the question of rural energy supply has to do with both availability and affordability and so we cannot ignore the effect of cost when assessing demand levels. The success of kerosene as an energy source for lighting rural homes will continue to rely on distribution packages that give access to small quantities as discussed above. Such methods will have to be introduced in electricity consumption if it is to be attractive to the low-income groups.

1.8 Global Energy Situation

The energy supply and consumption, looked at from a global perspective, presents a very interesting irregular pattern that obviously appears to favour certain regions of the world. The major global energy sources are oil, coal, natural gas, nuclear, hydro, geothermal, solar, wind and biomass while the commonly required end-use energy forms are electricity and heat. Of the two forms, electricity is the more versatile form as it can supply heat, light and also power industrial machines and equipment as well as household appliances. All this can be effected at the end-user stage. Heat energy, on the other hand, cannot do all these without the introduction of complicated conversion devices, which cannot be done by the end-user. Oil is used mainly in the transport sector. Practically all energy sources can be used to generate electricity but the cost would vary with the type of source. Taking this into consideration and the required generation capacity, electricity generation from nuclear, hydro and geothermal energies is more economically viable than from other sources. The next viable but slightly more expensive electricity is obtained from oil, coal and natural gas. All these suitable materials for electricity production occur naturally on earth and are not equitably distributed among nations of the world. Those countries that are not endowed with resources such as oil, coal, natural gas and nuclear substances have to buy them from the lucky nations. Hydro and geothermal sources are not transferable and therefore there is nothing that countries without them can do. All nations can rightly claim to have both sunlight and wind even though in varying quantities. The situation implies that rich nations are in a better position to acquire all their energy requirements even if they are not naturally endowed with many of these suitable resources. Poor developing nations have only two reasonable options: development of energy conversion technologies for locally available renewable energy resources; and, purchase of energy resources and materials from other nations. The achievement of any or both of these options requires a combination of resources, strong economy and good resource management, all of which are in short supply in the developing countries. As a result of this, over two billion people in the world have no access to electricity and most of these are in the developing countries especially sub-Sahara Africa. The global campaign on the need to protect the environment is not making the situation any better because energy production and consumption is one of the largest sources of environmental pollutants. There are also restrictions on the development of nuclear technologies for both security and environmental reasons. But this notwithstanding, large-scale electricity generation from the identified sources are just too expensive for the poor economies of most developing nations. This situation appears to rule out the prospects of eventually making electricity available to the over two billion people, which means they will continue to be shut out from the rest of the world because they will not be able to use modern technologies such as electronic communication systems which have made the world to become a small village. This global village will never be complete without the isolated two billion brothers and sisters who are languishing in poverty and darkness. The result is that the per capita energy consumption in the poor regions will always be well below that of the industrialized world and any effort to supply electricity in these countries will continue to be hampered by the state of poverty. There is also no indication at present that electricity will one day be supplied free of charge to the poor communities. So we see here a vicious cycle that will be difficult to break. But this does not mean that there is no hope for the future. There are alternatives, which can be developed to suit the circumstances of the rural poor and these can only succeed if the governments of the poor nations can implement deliberate initiatives that seriously address the unique socio-cultural and economic situation prevailing in the rural areas. One interesting observation is that in the industrialized nations where there are several options for generating electricity that include nuclear power, large hydro potentials and home-based fossil fuels, interest in the development of more sources is growing. The uses of fossil-based fuels and nuclear power make very significant contributions in the generation of electricity but new development of such facilities has drastically reduced. Even investment on new large hydro facilities has also remarkably decreased. The United States which is one of the world's largest power consumers has already recognized the enormous potential in renewable energies such as solar, wind, ethanol or bio-diesel and hydrogen fuels as energy sources that could play a significant role in future and is encouraging intensive research and developments even though they are considered to be more costly than the traditional energy sources. This is part of the government's strategy of reducing dependency on foreign oil in order to secure economic and national security. The plan also has the added advantage of slashing the pollutants and greenhouse gas emissions. Hydrogen powered vehicles are already in use in numerous pilot projects around the world. The technology itself has been proven to be viable but it is still costly for general use and the method for distribution of hydrogen to consumers has not been established. In Europe, investment on renewable energy technologies is continuously increasing especially wind power, which is considered to be the fastest growing segment of the global energy industry. There are plans to invest over USD 3 billion to set up wind farms in Europe. One of the sites that has been identified is the estuary of the Thames river, off the British coast, where a wind farm with a capacity to generate about 1,000MW could be established. Such a plan, if successful, will also reduce greenhouse CO_2 emissions by up to about 2 million tonnes a year. Other renewable energies (hydrogen, bio-fuels and solar) are also under serious considerations. It is envisaged that the renewable energies will contribute between a quarter and a third of the world's energy mix by 2050 and companies are working on strategies for benefiting from such large business opportunities. Of the current 50,000MW of wind power installed around the world, 34,000MW are in Europe. With the forecast growth, it is expected that about 100,000MW of wind power, involving billions of dollars in investment, will be operational by 2010. So far Europe has dominated wind energy but there is a general global shift to both wind and solar and countries like China and India are also fast-growing renewable energy zones. In the African continent where there is great potential for solar, wind and bio-fuels, the debate on their development is, at best, low-keyed. It appears, in terms of renewable energy development, Africa has already accepted to trail behind the rest of the world even though it had the opportunity to take the lead in some of these technologies.

1.9 Concluding Remarks

This chapter has given some important information about East Africa and has high-lighted existing energy management practices and choice factors with regard to commercial and non-commercial energy resources. The east African energy use pattern is compared with the global energy consumption pattern. Further relevant information is obtainable from [8, 9, 11, 20, 44, 50, 53, 54, 55, 57, 60, 61].

2

East African Energy Resources

2.1 Introductory Remarks

Although there is no proven exploitable oil deposits in East Africa, the region is endowed with a variety of other energy resources that have not been fully exploited. There are a number of rivers, small and large, whose hydro-generating potentials are yet to be fully utilized. Deposits of coal, natural gas and geothermal sites have been identified and partial exploitation is in progress. Availability of biomass materials, which are a major source of heat energy, is an equally important aspect that must be considered. The region has a rich biodiversity in both flora and fauna. The vegetation cover includes tropical forests, savannah type of vegetation and arid and semi-arid characteristics. There are also mountainous regions such as Mt.Kilimanjaro, Mt. Kenya, and Mt. Elgon regions that experience fairly cold climate and high rainfall. On average, East Africa receives annual rainfall of about 1,500 mm, which is sufficient for constant regeneration of vegetation, and therefore there is a huge biomass potential. Since the region straddles the equator, there is abundant direct sunshine generally rising to a peak power of about 800Wm^{-2} as shown in Fig. 2.1. There are minor variations to this sunshine pattern especially in the months of November, December and January and also May, June and July when the sun is overhead near the tropics of Capricorn and Cancer respectively.

Wind energy is more site-specific than solar, since there is a minimum wind speed limit for wind turbine operations. And so, as much as there is wind everywhere, its speed in some places may not meet the minimum required level for the operation of a wind generator of desired capacity. In East Africa, particularly in Kenya, the altitude rises

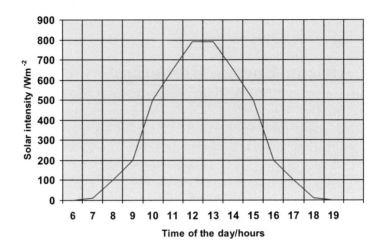

Fig. 2.1. Hypothetical average normal clear sky solar radiation in East Africa

rapidly from sea level on the east coast to over 1500 m above sea level in the central parts of the country. There are various highlands in the region where wind speeds are sufficiently high and therefore suitable for application of wind energy. It has been established that wind power generation potential of about 345 Wm^{-2} is available in Nairobi, Eastern, North Eastern and Coast provinces of Kenya. Lake Victoria region, the highlands and northern part of Kenya also experience regular wind speeds that can operate wind machines. All these energy sources (hydro, geothermal, biomass, natural gas, coal, solar and wind) are considered as well as their potential applications. Although fossil-based fuels like petrol, diesel, LPG and kerosene are making very significant contributions in the region's economic development especially in the transport sector and rural lighting; they will not be discussed in details because all of them are imported into the region. The importation and distribution of oil-based fuels are handled almost exclusively by the private sector with a lot of support and keen interest from national governments. The governments' interests include making sure that there is enough oil in the three countries at all times and they can do this by providing and monitoring oil storage facilities and import requisition documents without directly getting involved in the trade. There are however some oil companies in which the governments have significant shares. On occasions when there is a crisis, the governments can get involved in direct importation of oil. In deed all the three East African governments have established and used statutory regulatory bodies for these purposes for many years. There are only two major energy resources for industrial and economic development of the region. These are hydropower facilities and imported fossil fuels. Hydroelectric power facilities produce electricity for both domestic and industrial applications while fossil fuels support the transport sector as well as small power generation facilities. There are other energy resources such as geothermal, natural gas and coal that supplement the two major commercial energy resources. Table 2.1 gives the distribution of commercial energies in the three states of East Africa. Biomass, which is a major source of energy, is not considered as a commercial energy and is therefore not included in Table 2.1.

Table 2.1. Commercial Energy distribution in East Africa

	Electricity (Installed capacity/MW)	Oil consumption (Barrels per day)	Gas reserves (billion m^3)	Coal (thousand tonnes)
Kenya	1030	57,000	0	0
Uganda	380	9,000	0	0
Tanzania	860	17,000	33 (Songo Songo) 21 (Mnazi Bay)	200(annual output)

It is clear that the only energy resources that East Africa can claim to possess are solar, wind, hydro, biomass energies and some small quantities of geothermal, coal and natural gas. Both coal and natural gas were recently discovered in Tanzania and have not made any significant contribution in the energy scenario in the region. More attention will therefore be focused on the potentials of solar, wind and small hydropower systems in the region. Details of these resources and their technologies as alternative sources of energy are given in Chapter five.

2.2 Natural Gas

Tanzania is the only country in East Africa where there is proven natural gas reserves with a possible potential of up to 30 billion cubic meters. One of the gas fields with reserves estimated to be about 20 billion cubic meters is located on Songo Songo Island in the Indian Ocean, south-east of the coastal commercial city and former capital Dar-es-Salaam. Another natural gas site with about 500 million cubic meters is at Mnazi Bay. The immediate plan for the gas was to use it as fuel for the existing oil-fired electricity generators in the country and up-grade some of them so that they can feed directly into the national grid network. Of special consideration was the thermal electricity generation in Dar-es-Salaam, particularly the relatively large Ubungo thermal generator (about 112 MW), which was to be converted to use the gas. In order to do this reliably and with minimum transport cost, about 340 km of gas pipeline is to be built to convey the gas directly to a depot in Dar-es-Salaam from where it would be distributed to various electricity generators. The deposit at Mnazi Bay was to be used in Mtwara town to generate about 15 MW of electricity as well as in Dar-es-Salaam. In addition, the gas was to be used to produce fertilizers for both domestic and export markets as well as liquid fuel substitution in industry and transport sectors. The Tanzanian government agreed with foreign companies through licensing arrangements to handle production of the gas. Normally the exploitation of resources of great values such as this one needs assistance from development partners who prefer a package that includes the participation of more experienced companies. During such negotiation, there emerged a disagreement between the Tanzanian government and a powerful international development partner over the logistical arrangements regarding production and distribution of the gas. This initially caused some delays in completing the planned developments but eventually it was accomplished. Both Kenya and Uganda have not been able to find any trace of recoverable natural gas within their territories. However, despite its availability in Tanzania, natural gas has not made any significant impact as an energy source. It appears that its use is limited to large-scale power generation and the purpose of this is to reduce national bill for oil importation.

2.3 Coal

Active search for coal has been going on in the region with some mixed expectations particularly in Kenya where efforts are focusing on Mui and Mutito areas in Mwingi and Kitui districts. In Tanzania, commercial coal production started at Kiwira coalmine in Mbeya in the late 1980s with an estimated annual output of about 150,000 tonnes of raw coal out of which about 93,00 tonnes of processed coal would be obtained. There is also coal deposit at Mchuchuma in South west Tanzania near the northern tip of lake Nyasa. Due to the poor quality of Tanzanian coal and the high cost of transportation, coal has not become a major source of energy for domestic application. This has slowed down coal output but it is expected that once coal production is fully developed in the region, it will be used to diversify electricity generation and also supply industrial heat energy requirements not only in Tanzania but also in Kenya and Uganda. It is however not clear whether coal can, in future, make any impact as domestic heat energy source in the region since space heating is not an essential requirement.

2.4 Geothermal

Geothermal is a rapidly developing energy source in the East Africa region. Energy is obtained from natural heat stored in rocks and water within the earth's crust. Since the energy is extracted by drilling wells to the underground pressurized hot water and steam reservoirs, the wells must be shallow enough for energy production to be economically justifiable. The steam is led through pipes and finally used to turn turbines, which drive electricity generators. Kenya is the first country in the region to utilize geothermal energy. The stations are located at Olkaria near the southern shores of lake Naivasha (see, e.g., Fig. 2.2 in p. 50). The first of the three steam turbines producing a total of 45 MW started operating in 1981 while the last one was commissioned in 1985. About 33 wells have been dug to supply steam to the plant. The second phase of geothermal energy development known as Olkaria II will produce 64 MW when completed. The third geothermal site, Olkaria III, developed and totally owned by an independent electricity generating company, was also expected to produce another 64 MW when completed. By the beginning of 2003, the new plant was already producing 13.5 MW and by 2005 it was completed, bringing to 109 MW the total installed capacity at Olkaria. This capacity is soon expected to increase to 173 MW when Olkaria II will be fully operational. The independent power producer owning and operating Olkaria III is the first private company to use air-cooled converters that ensure zero surface discharge. This is a new and most environmentally benign technology in the electricity generation in Kenya. The company sells its electricity to Kenya Power and Lighting Company (KPLC) for nation-wide distribution. But, even

Fig. 2.2. Olkaria Geothermal Power Station in Kenya

with all these new power plants, geothermal development is still less than 10% of the estimated geothermal potential in Kenya, which stands at about 2,000MW. When eventually all the planned power plants will be operational, geothermal will contribute about 16% to 20% of the total electricity supply in the country. The Great Rift Valley, which is known to have favourable geological characteristics for geothermal energy, divides Kenya into almost two equal parts and the Kenya government is continuing to search for more geothermal sites. It is expected that more independent power producers will participate in the development of geothermal electricity production. In Uganda, estimates made several years ago put geothermal potential at about 450 MW located in Uganda's rift valley region but since then a lot of research has been carried out to determine if there is more geothermal potential in the country, particularly in the same Rift Valley which lies along its boarder with the Democratic Republic of Congo. The search for geothermal sites has kept alive Uganda government's hope of diversifying its electricity generation from the predominantly hydro sources. The areas that, according to geological properties, have shown some prospects include Katwe, Kibiro and Buranga but other areas bordering the rift valley in south-west and northern parts of Uganda also have volcanic and tectonic features that are indications of possible geothermal occurrence. It has been established that suitable underground rock temperatures ranging from 120° to 200°C are found in these areas. But, so far, Uganda is yet to exploit the identified geothermal potential. In Tanzania, very little effort has been put on the search for geothermal resources despite the fact that the central portion of Tanzania is in the Great Rift Valley, which is known to have geological formations that are characteristics of geothermal resources. The only exploitation of geothermal resources in the region is done in the Kenyan portion of the same Great Rift Valley.

2.5 Petroleum-Based Fuels

Deposits of significant quantities of oils have not been discovered in the region but each national government has put a lot of efforts and investments in the exploration of oil. This is done under government corporations specifically established to oversee activities in the petroleum industry. For example in Tanzania Petroleum Development Corporation (TPDC) is responsible for all aspects of petroleum industry including exploration, production, refining, storage and distribution. In Kenya, it is the Kenya National Oil Corporation that is responsible for this while Kenya Pipeline Company (KPC) deals with the distribution of all oil-based fuels to major storage facilities. Private sector involvement is also very strong in the petroleum distribution sub-sector and is the main supplier of oil fuels to the end users through region-wide network of petrol/gas stations. Most of these were multinational oil corporations but have, over the years, sold majority shares to local companies in order to consolidate control of local energy issues. There are more than fifteen different oil distributing companies operating in the region with the most prominent being MOBIL, BP, CALTEX, SHELL, TOTAL, ELF, FINA, KOBIL and AGIP. New players have recently joined the market mainly from South Africa, and are spreading their networks to cover the whole region. Many of them prefer to enter into the region through Ugandan and Tanzanian markets where competition is not as stiff as in Kenya. Oil-based fuels of various grades (diesel and petrol) are the major sources of energy for the transport sector while kerosene, which is another oil-based fuel, is the main source of power for lighting especially in the rural areas and among the urban low and middle income groups.

2.6 Hydropower

Hydropower is the main source of electric energy in the region and although more than 90% of the rural population has no access to electricity, the industrial production heavily depends on it for a wide range of applications. Electricity therefore can be considered to be the main driving force in industrialization process where it plays a pivotal role while oil sustains the transport sector. Both oil and electricity are sources of energy that are of extreme economic significance for any nation. More often than not, when talking about national energy plans and strategies in developing countries, the reference obviously is on oil and electricity. In fact in the East African region the word power means electricity while energy or fuel means oil. Although electricity accounts, on average, for less than 4% of the total energy consumption in the region, while oil contributes less than 20%, they are nevertheless the most important energy commodities for the national governments. Oil is imported into the region and does not require any special development plan but hydropower is internally generated and therefore needs careful production projections. Consequently, it is important to analyze the electricity sub-sector in details including the potential future role of hydro resources in rural energy supply strategies. In Kenya, electricity is produced from hydro, geothermal and fuel-operated generators. The present installed generation capacity is about 1085MW but most facilities are operating below capacity due to a number of reasons including occasional low level of water in the hydro dams as a result of prolonged drought and so the effective capacity is only 1032MW. Peak power demand however, is low and is about 890MW on average. The distribution is under the sole monopoly of the government-controlled Kenya Power and Lighting Company (KPLC). Hydropower however is not the only source of electricity. There are also other electricity sources such as co-generated power and some oil-fired generators, which are not included in the quoted installed capacity because many of them are not available for public consumption. If all sources were considered then this capacity would be about 1200 MW. However, there are variations to this so that the capacity sometimes goes way below the demand due to a number of factors, forcing the distributor to ration electricity. A year of little rain than expected is considered to be a year of drought. The problem of increased silt in the dams is also experienced during too much rain reducing the amount of water and hence less electricity generation. One very important aspect of the Kenyan hydro electricity generation is that more than half its installed capacity is generated from one river, the Tana river (see, e.g., Table 2.2). The major power stations on river Tana are Masinga, Kamburu, Gitaru, Kindaruma and Kiambere and are known collectively as the Seven Forks Hydro Stations generating a total of 563 MW, which is about 60% of electricity consumed in Kenya. The stations are not only along one river but they are also linked together by cascading water from one station to another taking advantage of the head pressure created by each dam. The dam at Masinga is used as the main reservoir that supplies the rest of the stations with water during the dry season. This linkage and their location on one river make all of them equally vulnerable to the effects of drought. The obvious result is that the country is likely to face occasional national electricity shortages when the river and its catchments experience dry spells even if there is enough rain in other parts of the country. In deed, this has been practically experienced a number of times forcing the power company to enter into hasty negotiations with independent and emergency power producers. In order to reduce the risk of power shortages, plans are underway to increase installed capacity by 392MW by the year 2008 by developing new plants as well as rehabilitating some of the old facilities. It is expected that Independent Power Producers will also have some contributions in these plans by injecting about 80MW by 2008 under 20 years power purchase agreement. A possibility of importing power from Tanzania through Arusha using 330KV transmission line to Nairobi is also under consideration mainly because Uganda is likely to stop power export to Kenya due to increased demand at home. However, it is most unlikely that these developments will reduce power costs or increase electricity access to the ordinary rural Kenyans as discussed in Chap. 8. Perhaps the only most important aspect of these arrangements is that they will diversify power sources from the Seven Forks system and increase security.

The linkages in the Seven Forks system are very interesting and might have been favoured because of low construction cost. In addition to cascading water from one station to another, generated electricity is also in certain cases conveyed to another station before transmission to Nairobi, the capital city. For example, electricity generated at Masinga is first transmitted to Kamburu before transmission to Nairobi. Similarly, electricity from Gitaru is first transmitted to Kamburu. This means that electricity from the Seven Fork system can only be transmitted to Nairobi from three points: Kamburu, Kindaruma and Kiambere.

Table 2.2. Electricity Generation in Kenya (Large hydro-stations)

Station	Location (District/ River)	Capacity (MW)	Year commissioned	Reservoir capacity (million m³)
Masinga	Tana	40	1981	1560
Kamburu	Tana	94.2	1974	Underground station served by Masinga dam
Gitaru	Tana	225	1999	Water from Kamburu through 2.9 km tunnel
Kindaruma	Tana	44	1968	Water from Gitaru through 5 km tunnel
Kiambere	Tana	144	1988	585 (receives water from Kindaruma
Turkwel Gorge	West Pokot	106	1991	1600
Total installed capacity from large hydro stations		653.2		

In addition to the large Seven Forks hydro stations and Turkwel Gorge, there are a number of small hydro stations some of which are now more than 50 years old and are still operating. The European settlers introduced many of the small hydropower machines in Kenya in the early and mid 20th century. A good number of them are no longer operating partly due to government's lack of interest in small hydro and partly because of aging and rising cost of maintenance. Table 2.3 gives the details of some stations that are still in action. Small hydro systems with power ranging from 400 to 800 KW have been operating in the tea-growing areas of Kericho highlands where some European settlers are still doing some farming. Tenwek Hospital in the same area is one such example where a small 400 KW hydropower facility is used to provide power for lighting. In general small hydros have been used in Kenya since 1919 and compared to other technologies under similar conditions of lack of spares and maintenance knowledge, the small hydros have done reasonably well. At least, some of them are still operational. A survey carried out on small hydro potential in Kenya revealed that there over 100 sites that are suitable for generation of more than 10 kW.

Table 2.3. Small Hydro power stations in Kenya

Station	Location (District/ River)	Capacity (MW)	Year commissioned
Mesco	Maragua	0.38	1919
Ndula	Thika	2	1924
Tana	Upper Tana	14.4	1940 (3 machines) (additional 2 machines in 1953)
Sagana	Upper Tana	1.5	1952
Gogo	Migori	2	1952
Sossiani	Sosiana	0.4	1955
Wanjii	Maragua	7.4	1955
Total installed capacity from small hydro stations		28.08	

The Tanzanian hydro electricity production is smaller than the Kenyan system but targets a larger area. All hydropower generation is still done by the state monopoly company, the Tanzanian Electric Supply Company Limited (TANESCO) but plans are at an advanced stage to change this. The total installed hydro capacity is 561 MW generated as shown in Table 2.4. Kidatu is the largest hydro station in Tanzania with an installed capacity of 204 MW followed by Kihansi with an installed capacity of 180 MW. The rest are producing less than 100 MW with Nyumba ya Mungu being the smallest at only 8 MW. A small amount of hydropower is imported from the neighboring countries such as Zambia (3 MW) and Uganda (10 MW). These are mainly used to serve the regions near the borders with these countries.

The Ugandan Electricity generation sub sector is the smallest in East Africa partly due to the fact that Uganda is the smallest in size and population in the region and partly due to stalled development programmes during the years of civil war of the early 1980s. The presence of river Nile with a large hydro potential has affected the development of small hydro power facilities but there are a few that were introduced by the colonial settlers and, more recently, by non-governmental organizations especially the church. Since the late 1980s when the country

Table 2.4. Hydro Electricity Production in Tanzania

Station	Installed Capacity (MW)
Kidatu	204
Kihansi	180
Mtera	80
Pangani	68
Hale	21
Nyumba ya Mungu	8
Total	561

returned to peaceful governance, a number of sites for small hydro development have been identified and plans are underway to develop them using local expertise and locally fabricated equipment. In the mid 1990s a survey was conducted in West Nile region and a total of about 80 sites with potential ranging from 2KW to about 600 KW were identified. As for the large hydropower, the Nalubale Power Station, which is located in the south-eastern part of Uganda at Owen Falls, on river Nile, is one of the oldest large electricity generating facilities in the region. It has had for a long time a generating capacity of about 180 MW but the recent extension that brought in the Kiira Plant has increased the capacity to about 300 MW. This extension was commissioned in 2001 and was expected to produce 200 MW but only 120 MW has been connected to the national grid network. The two stations on river Nile near the town of Jinja are the only major sources of hydropower in Uganda. Again the management of the electricity sub-sector has been done by the Uganda Electricity Board in a similar monopolistic approach as were in Kenya and Tanzania but the recently introduced reforms are set to change this in the whole region.

2.7 Thermal Generators

Being the largest country in the region, Tanzania has experienced more serious electricity distribution problems than the other sister states. As a result of this, the country chose to use thermal generators to power its scattered towns and villages. Some of these are connected to the national grid lines while others are isolated facilities. Those that are connected to the national grid are located in the larger towns and cities and therefore mostly found in Dar-es-Salaam, Mwanza, Tabora, Dodoma, Musoma and Mbeya. The total generation capacity is estimated to be about 80 MW but have been producing just about 50% of this due to operational problems. The isolated thermal generators are meant to serve low populated remote settlements. Such generators are located in Mtwara, Kigoma, Njombe, Lindi, Tunduru, Mafia, Mpanda, Ikwiriri, Liwale, Songea, Kilwa Masoko and Masai and have a total generation capacity of 31 MW. The Tanzanian case for these thermal generators is quite unique and was encouraged by the socialist economic policy that Tanzanian government pursued for many years after independence in 1961. The policy encouraged people to get together and live in large communal villages and the government was obliged to support communities' efforts to acquire the basic facilities and this is how small townships got their generators. Many of these generators are facing myriad problems due to old age. In addition to these old thermal generators, the independent power producers that have recently entered into the country's energy sector find it more economically convenient to use thermal generators. Consequently, one such power producer, Independent Power Tanzania Ltd (ITPL) is already producing 100 MW, which it sells to TANESCO for distribution. Another independent producer, SONGAS, with a capacity of 200 MW is already supplying 120 MW. This particular company is using the available natural gas in Tanzania for electricity generation. The other independent power companies are the Tanganyika Wattle Company (TANWAT) with a capacity of 2.5 MW and Kiwira Coal Mine with a capacity of 6 MW. Both of them are using thermal generators. Like Tanzania, Kenya has a number of thermal generators. The majority of them are managed by the Kenya electricity Generation Company (KEGEN), which sells power to the distribution company, Kenya Power and Lighting Company (KPLC). The largest of these is the Kipevu system located at the Indian Ocean coastal city of Mombasa. The system uses three different methods of generation: thermal, Diesel engines and gas turbines. The first Kipevu thermal machine was commissioned in 1955 and was gradually up-graded to seven machines by 1976. These were oil-fired drum type boilers. The first five machines have been retired due to aging while the last two machines are still operating with a total capacity of 63 MW. The first Kipevu gas turbine was commissioned in 1987 with an installed capacity of 31 MW while the second one with a capacity of 32 MW was commissioned in 1999, bringing the total gas capacity to 63 MW. The Kipevu diesel generator was commissioned in 1999 with an installed capacity of 73 MW. Up in Nairobi, there is yet another gas turbine, which was commissioned in1972 with an installed capacity

of 13.5 MW at Nairobi South. All these are connected to the national
grid network but there are also three isolated thermal generators serving remotely isolated towns. These are the Garissa Power Plant with
an installed capacity of 2.4 MW and supplies power to Garissa town in
the remote part of North Eastern Kenya; the Lamu Power Plant that
supplies electricity to the traditionally unique Lamu Island; and the
Marsabit thermal generator. Marsabit is an isolated town in the northern part of Kenya where security risks make it difficult to manage grid
power lines. There are a few independent power producers using thermal generators but their impact in the relatively large Kenyan power
sub-sector is yet to be felt. One such producer is the Tsavo Power Ltd,
which has an installed capacity of 74 MW. There are also other thermal
power generators in Lanet (55 MW), Eldoret (55 MW), Embakasi (105
MW) and Ruaraka (105 MW).All power producers in Kenya, including
the government-controlled KNGEN, sell power to the sole distribution
monopoly KPLC. The other state in East Africa, Uganda, gets most
of its electricity from hydro power stations. Electricity from thermal
systems is less than 5 MW. This could be so because of the fact that
Uganda is a small country compared to both Tanzania and Kenya and
therefore distribution of electricity from centralized hydro stations may
not face as much problems as those faced by Kenya and Tanzania. Furthermore electricity production in Uganda is much lower than in the
other two countries. The recent reforms in the power sector are expected to attract some independent power producers who generally
prefer to use thermal generation methods. When this happens, then
Uganda is likely to increase its capacity in thermal electricity production. The present political will to bring the three East African states
closer together may encourage more thermal generators based on coal
and natural gas produced in Tanzania.

2.8 Co-Generation

Co-generation is a system in which both electricity and some other
product such as process heat are produced from the same power plant.
It is also often considered as a process in which electricity is generated
as a secondary commodity in addition to the core product of the establishment. For example, timber mills need electricity to operate while at
the same time they produce a lot of wood waste that can be used to
produce that electricity. Similarly, sugar factories need electricity but,

to produce sugar, they need a lot of heat for their boilers, and this heat is obtained by burning sugarcane waste (bagasse) and the steam from the boilers can be used to turn turbines for electricity generation. Thus their core product is sugar but they find it also convenient to generate electricity for their own use since steam is available. This technology is globally proven and indeed many of these factories have been doing just that for a long time but the laws governing the power sector did not allow them to produce electricity as a commercial commodity. This has been a very unfortunate situation in countries where surrounding communities still rely almost entirely on biomass as the source of energy. Thus co-generation, attractive and cost-effective as it may be, has not been exploited in East Africa for the benefit of rural people. In Tanzania, co-generation has been practiced at sawmills such as Sao Hill and Tanganyika Wattle Company and also at sugar processing plants such as Tanganyika Planting Company Ltd (TPC), Kilombero Sugar Company, Mtibwa Sugar Estates and the Kagera Sugar Company. A few of these like the Tanganyika Wattle Company (TANWAT) are already making use of the reforms to sell power to Tanzania Electricity Supply Company (TANESCO) for national consumption. There is, however, a growing interest in producing electricity as part of the commercial outputs with a view to selling it in order to improve economic performance. In Kenya, there are also a number of sugar milling factories and other agro-based companies, which have the capacity to increase their co-generation levels but have not made any contribution in the supply of electricity to the public. These are Sony Sugar, Nzoia Sugar, Mumias Sugar, Chemelil Sugar, Muhoroni Sugar, Agrochemical and Food Company and the Kisumu Molases Plant. All these companies are located in western part of Kenya and are capable of supplying the region with a substantial amount of energy. They are all close to Lake Victoria and together with Kagera Sugar Company in Tanzania and Kakira Sugar Company in Uganda, which are also close to the lake. could effectively serve much of the lake Victoria region. Already Kakira had prepared to increase its electricity capacity to about 15 MW or more with the aim of selling the surplus to the Uganda Electricity Board. The desire by these companies to actively co-generate electricity for commercial distribution is a progressive development that should be encouraged by the respective East African states. The seven sugar milling companies in Kenya produce an average of about 1.8 million tones of bagasse. Out of this, about 56% is used in co-generation with a total installed capacity of 25 MW which is all consumed in-house. The rest of the bagasse is disposed at a cost to the company but with some incentive and appropriate energy policy, this can be used to generate more electricity that would bring additional income to the company and also increase national electricity generation capacity.

2.9 Solar and Wind Energies

Solar energy can directly provide heat and electrical energies while kinetic energy of the wind can be converted to electrical energy through a generator and also to mechanical energy through appropriate mechanical linkages. All these final forms of energy are in high demand in any modern society. The conversion technologies for both solar and wind energies are well developed and their efficacies and suitability have been proven worldwide. East Africa straddles the equator and therefore has adequate amount of sunshine throughout the year for solar energy conversion devices. It also has geographical features and climatic conditions with wind speeds that can be converted into useful energy. These have not been adequately exploited in the region and the blame goes to national energy policies and implementation strategies that have not given these energy resources the attention they deserve. Although they are covered in energy policy documents their implementation is often limited to tax reductions given to those importing them into the region or end up simply as political rhetoric. Over the years, a local capacity to fabricate solar water heaters has been developed but there has been very little support for promotion. Such capacity exists in both Kenya and Tanzania where locally fabricated solar water heaters are used by both individual customers and institutions such as hospitals and hotels. Bugando Hospital in Mwanza, Makiungu Hospital in Singida, Morogoro Hotel, Hotel 77 in Arusha, kilimanjaro Christian Medical Centre, Manyoni Mission in Singida are a few examples of working solar thermal systems. In total, there are over 600 solar water heaters installed and working in Tanzania. There are also some photovoltaic solar systems for electricity generation installed and working in the region. A few wind turbines are used mainly for water pumping in the rural areas. One notable wind generator is at Chunya Catholic Mission and provides 400 W of electricity to the mission mainly for lighting. Kenya is the leading country in the region with a large number of both solar water heaters and photovoltaic panels installed and working. Institutions such as hospitals, schools and hotels use the large ones while smaller ones are serving individual families in peri-urban and rural areas. It is estimated that over 50,000 photovoltaic panels of various power ratings are installed and operating in the rural areas of Kenya. Some large national corporations such as Posts and Telecommunication Corporations are using solar photovoltaic panels in their communication systems in order to enhance national coverage to include areas where there is no national grid electricity supply. National organizations in charge of wildlife protection are also using them to electrify fences in order to restrict the animals to the designated game parks. Kenya has a well-developed market for the devices but, again, the official support is limited to tax concessions. Two wind turbine manufacturers are operating in Kenya and most of their turbines are used for water pumping. However, the Kenya Electricity Generation Company (KENGEN) is operating two wind turbines with a total capacity of 350 KW in Ngong area near Nairobi. The two machines were imported into the country through assistance from a European country with more experience in operating large wind generators. The operation of these two wind turbines has proved that there is enough wind in the region for these machines. A third wind generator also serves the remote northern town of Marsabit but this one is unique because it is the only hybrid diesel/wind turbine system in the country. It is rated at 200 KW. These are the only known operational wind machines that generate electricity in Kenya. However, good potential for wind machines exist because Kenya has high wind speeds and there are two local and reliable wind turbine manufacturers that can provide after-sales services including spare parts. These manufacturers have locally produced and sold several wind machines mostly for water pumping in various parts of the country. Over 50% of Kenya has an average annual wind speed above 3 ms^{-1}. In Tanzania, a lot more local efforts have been put on the development of wind machines for water pumping but relatively little on wind generators. For example in the 1980s the government supported the installation of wind pumps to supply water to several villages especially in Singida region. Several workshops were also established with support from the government to encourage the production of prototype wind pumps. Such workshops were at Ubungo, Faculty of Engineering at the University of Dar-es-Salaam and Arusha Appropriate Technology Project, (now part of Centre for Agricultural Mechanization and Rural Technology). Some very good progress were made and machines produced and tested but their high cost prohibited their widespread use in the country. In Uganda the situation is a little different since the government has not provided significant support for the development of wind machines. Those that are operating were installed by Non-governmental organizations such as the Roman Catholic Mission and the Church of Uganda. One government institution, the Karamoja Development Authority, installed some wind pumps but this is an isolated case in Uganda. These applications of wind machines in the region have demonstrated clearly that the region has the potential for wind energy in terms of both their application and local production and there are many areas with adequate wind speeds for electricity generation.

2.10 Biomass Energy

Biomass energy resources cover a wide range of materials that include gases, liquids and solids such as biogas, producer gas, ethanol, charcoal and wood. In urban slums, peri-urban and rural areas of East Africa, wood is the main source of energy and therefore it will be discussed in details even though it is of little national economic importance. For a long time, it was so easily available in the immediate neighbourhood that it could not be imagined that one could buy it for use as an energy source. Today, much of it has disappeared and wood fuel is now a commodity that one can buy in the local markets even in the rural areas. It is, however, still very cheap compared to other energy resources and most people can still obtain it without having to buy it. It is also a very interesting energy source and many energy specialists still find it difficult to quantify wood in terms of energy quantity since it does not have standards or set properties for accurate determination of its energy properties. In addition to variations of wood species, wood can have different energy values depending on it moisture content. Although water in the wood may not have heat value, it nevertheless reduces the net heat value of the unit weight of the wood since it has heat capacity and latent heat of evaporation that must be supplied from energy in the wood when the temperature is raised to a combustion level. So the energy of the wood falls rapidly as moisture content increases. Table 2.5 gives examples of variation of heat content in wood as a function of moisture content. Wood is also bulky with weights that depend on the moisture content and therefore its transport cost is generally high compared to the amount of energy obtained from it. Charcoal, which is produced from wood, contains about 50% of the energy in the original wood, weighs only about 25% and has more consistent properties than wood. Its calorific value is comparable to those of some industrial grades of coal. These qualities make it a popular source of energy for low-income urban population who do not have much storage and cooking space. In addition to its high energy content, charcoal also has the advantage that it does not burn with much flame that could cause any danger or discomfort to the user; it simply glows and produces the required heat.

Table 2.5. Energy values of wood and wood products

	Moisture content	Energy content (MJ kg^{-1})
Wood	100 - 120%	8.3
	15 - 20%	16.4
	8 - 10% (bone dry)	19.3
Charcoal	5 - 10%	29
Wood gas	-	7

The level of biomass contribution to the total national energy consumption in Kenya has gone down to about 70% from about 82% level of the 1980s. In Uganda and Tanzania, it is still contributing about 90% of the total energy consumption. This is a direct indication that the urban population in the region is still a very small fraction of the total population. The low consumption level of biomass in Kenya could be attributed to the fast growing number of the so-called rural market centers where small traders settle to carry out their businesses. Residents of such centers tend to rely on kerosene for both cooking and lighting. As discussed in chapter one, these centers also normally spring up faster along national grid power lines. So some successful traders use electricity and LPG as their domestic energy sources. The main factor, however, is the high growth rate of urban population of young people in search of job opportunities. It is now estimated that about 30% of Kenya's population is urbanized. Kenya is also more economically developed than either Tanzania or Uganda and this explains the relatively higher per capita clean energy consumption and lower biomass use compared to Tanzania and Uganda. It is nevertheless clear from Tables 1.5, 1.6 and 1.7 that almost the entire rural population in East Africa depends exclusively on biomass energy resources. This also implies that although it is becoming more and more difficult to obtain, biomass is still available in sufficient quantities to satisfy the demand. Many people are aware of its crucial value as an energy source and are making efforts to plant various types of trees for this purpose while at the same time using plant residues as energy source. Individual forestry master plans for Kenya, Uganda and Tanzania indicate that wood resources are on the increase in farms while they are decreasing in other land categories. This shows that the ordinary people are aware of the importance of biomass regeneration to the extent that tree planting is a readily accepted practice in the farms. The popular tree species planted in the farms include Grevillea robusta, Eucalyptus species, Cypress species and a variety of fruit trees.

2.11 Concluding Remarks

In this chapter, East Africa's prominent energy resources have been identified. A number of these are locally available while others are imported into the region. The three East African states of Kenya, Uganda and Tanzania are all endowed with renewable energies but there are variations with regard to the quantities and availabilities of some energies such as geothermal, coal and natural gas. Detailed information is available in [8, 11, 17, 20, 32, 43, 45, 53, 54, 55].

3

Energy Potentials

3.1 Introductory Remarks

Energy consumption in many developing countries is not clearly documented as people tend to use whatever they can get readily and cheaply and therefore both cost and the source become the most important consideration. In East Africa, the following are the common energy sources:

Firewood: This is a common source of energy used by about 90% of the rural households and about 10% of urban households. About 80% of these people obtain their firewood free while others either regularly purchase it or supplement their free collection by purchasing some. Firewood is mainly used for cooking and space heating.

Charcoal: Of those who use charcoal, 80% of them are urban households while less than 20% are rural households. The average per capita charcoal consumption in the region is about 150 kg, giving charcoal a significant trade volume, which, in financial terms, could be as high as 50% of the cost of oil imports.

Wood waste: Use of wood waste is gradually declining with less than 3% using it in the 1990s compared to the reported over 5% in the 1980s. This is due to the declining number of special stoves that use wood waste such as sawdust.

Farm residues: These are used mainly in the rural areas and are generally seasonal depending on the periods of harvest when they are available in large quantities. However their continued use will compromise opportunities for improving soil fertility.

Biogas: The contribution of biogas as an energy source both in rural and urban areas is negligible but the potential exists in some parts of the region where keeping animals is a traditional pride. This practice is however rapidly spreading to other areas where such traditions did not exist and there are already a number of biogas plants operated by some farmers and institutions. The potential for biogas will therefore continue to increase.

Kerosene: About 95% of the rural households and about 90% of the urban households use kerosene principally for lighting. It is estimated that the per capita kerosene consumption varies from about 40 litres in the rural areas to about 90 litres in urban areas. Quite a large number of urban households use kerosene for both lighting and cooking and this is one of the reasons for its high consumption in the towns.

Liquefied Petroleum Gas (LPG): This is usually used as a stand-by or emergency energy source and therefore very little is consumed. In urban households it is used as support for the more expensive electricity or when kerosene cannot be immediately obtained while in rural households it is used along with firewood. Using LPG has also been limited by the fact that distributors have special components that cannot be interchanged, for example, Total Oil Company uses LPG cylinder with a regulator that cannot be used by those who have regulators from other distributors.

Electricity: Electricity is the most modern and convenient energy form. It is also considered to be clean and versatile in its applications including the running of household appliances such as radios, TVs, refrigerators etc. It is however, expensive for the majority of households. Even in the urban areas where it is readily available, still less than 50% of the households use it regularly. The situation is far worse in the rural areas where it is not accessible in many areas. There are doubts that even if it were easily accessible, many people would not use it. The scenario in the towns is a clear evidence of this assertion that electricity may be accessible but still very few households would be connected. Almost all the electricity in the region comes from government controlled facilities with about 60% generated by large hydro power stations while others are from geothermal, oil or gas-fired generators. Renewable sources of electricity such as wind, solar and co-generation have not made any impact in the energy scene.

It is clear that there is a whole range of energy options but unfortunately some of these have not been adequately considered in official energy concerns even though they have been harnessed in East Africa for different purposes and at different levels in terms of applied technology. It is important, for long term planning, to also assess how much energy is available in the region for further development. This information enables energy planners and those who wish to invest in the energy sector to make decisions on the type of energy resource that would be worthy of consideration with respect to availability, consumption trends and quantity. We have noted that the end-user requires energy mainly in the form of heat, light or electricity for operating household appliances, equipment and machinery. As has been outlined, the sources of these energies in East Africa are hydro, fuel-operated generators, geothermal, oil, natural gas, LPG, coal, wind, solar and biomass. In addition to these, there is widespread use of rechargeable lead acid batteries and dry cells particularly for emergency lighting and home entertainment. Some of these are relatively more expensive but are used even by those considered to be poor. The reasons for this will be addressed later.

The discussion about the potential of certain energies may not be of much use in the case of East Africa since they are not locally available and their acquisition depends on how well the economy of the country is managed. In this regard the potential of oil as an energy source will not be discussed. The search for it, however, is in progress in the region especially in Kenya and Tanzania and some positive indicators have been identified but it is not clear whether economically recoverable oil will be found. Similarly, Uganda and Tanzania have not discovered any exploitable geothermal resources but some studies and research are being conducted to determine the possibility of geothermal occurrence in these countries. Other resources like natural gas and coal are also rare with Tanzania being the only country in the region endowed with some natural gas and coal. The potentials of such resources that are country-specific are discussed in relation to those countries. It is however important to note that if the current trend of oil consumption continues then East Africa will need more and more oil as the population increases and transport and related sectors expand. This will mean increased expenditure on oil imports further reducing the region's ability to support other necessary development projects. Therefore although imported fuels will not be discussed in details, it should be noted that increased demand for them will retard overall national development process. For this reason and in addition to their finite nature, oil consumption should be suppressed by developing attractive packages to attract consumers to alternative energy sources.

3.2 Hydro Energy

Lake Victoria, which is the source of river Nile is located almost at the center of East Africa and is sometimes referred to as the heart of Africa. River Nile is the only major river flowing out of the lake while several rivers flow into the lake especially from the Kenyan and Tanzanian sides. River Nile traverses Uganda before proceeding on to Egypt through the vast Sudanese territory where it is joined by the Blue Nile from the Ethiopian highlands. It is one of the two longest rivers in the world with an estimated length of about 5580 km from lake Victoria (Uganda) to the Mediterranean Sea (Egypt). With a high volume flow rate of about $600m^3s^{-1}$, river Nile gives Uganda a huge hydro power potential of about 2000 MW most of which has not been exploited. Uganda is further favoured by the fact that river Nile flows through the central part of the country making it possible to develop economically viable centralized generating facilities capable of electrifying the whole country. Despite the favourable distribution of her hydro potential, Uganda has only exploited less than 20% of this and is the least electrified state in East Africa. However, for a small country like Uganda, the construction of large hydropower dams would have a great impact on the environment and displacement of people and so the development of hydropower should be carefully planned. Some of the identified potential hydro power stations are shown in Table 3.1. The extensions of Owen Falls (Kiira)(see Fig. 3.2) and other power

Table 3.1. Hydropower potential in Uganda

Station	Estimated Potential Output (MW)
Ayago	580
Murchison Fall (Kabalega)	480
Kalagala	450
Bujagali	320
Karuma (kamdini)	200
Estimated Total Output	2,030

stations are not included in Table 3.1 since they are already operating but they also have potential for expansion to larger capacities. Some of the sites listed on Table 3.1 are already under serious consideration for development against stiff opposition from Ugandan environmentalists, particularly the Bujagali site (see Fig. 3.1), which is considered to be an important cultural and ecological site. Most of the hydropower potentials in Uganda are based along river Nile. There is also potential existing in other river regimes for small and medium power generation as shown in Table 3.2.

Fig. 3.1. The controversial Bujagali Falls under hydro station development

In total, over 20 small and medium hydropower sites have been identified in Uganda and since they are scattered in remote areas far from the main centralized large hydro site near the Lake Victoria port town of Jinja, they provide the country with a good opportunity for local independent power companies to produce electricity for isolated rural settlements and institutions such as schools and hospitals. The Uganda government has already put together a number of mini hydro projects that are expected to inject about 30MW into the grid. They include West Nile Hydro Project (5.5MW), Rukungiri-Bushenyi Power Project (4.6MW), Kaseses Rural Electrification Project (5 MW) and the Buseruka Hydro Power Project (15.6MW). Since Uganda covers small territory compared to Kenya and Tanzania, a single national

Fig. 3.2. Hydro power station at Owen Falls on river Nile in Uganda

grid network may serve her more effectively. Like Uganda, Kenya's hydropower generation is dependent on one river with five power stations connected together within the Seven Forks system on river Tana. The same river still has some undeveloped potential for more electricity generation at Mutonga, Grand Falls, Adamson's Falls and Kora Falls all of which have been identified for future development. The river, however, flows through a section of the country, which has relatively low population density, and power has to be transported long distances to reach the major towns and cities such Kisumu, Mombasa and the capital city Nairobi. Apart from river Tana, there are a number of smaller but permanent rivers flowing from the western ridge of the Great Rift Valley into lake Victoria. These rivers have a great potential for small and medium hydropower facilities. A few, such as Magwagwa and Sondu Miriu on river Miriu, Leshota, Oldoriko and Oleturat on river Ewaso Nyiro have been identified and targeted for development by KENGEN but there are still about 100 sites with smaller hydro potentials on rivers like Yala, Athi, Mara and Turkwel systems. Within the tea estates of Kericho highlands, there are small rivers with potential for small hy-

Table 3.2. Small hydro potential in Uganda

River	Estimated Potential (MW)
Anyau	0.3
Bunyonyi	1.3
Nyakabuguka	0.2
Mpanga	0.4
Nyakabale	0.1
Ataki	0.2
Kisiizi	0.2
Kagera	2
Nyagak	3
Kaku	2.2
Maziba	0.5
Ruimi	1.5
Ishasha	6
Sagahi	6
Mubuku	7.5
Muzizi	11
Mtungu	12
Total estimated for small and medium hydro generation	54.4

dropower production and indeed the viability of the systems have been proven by some institutions in the area that produce power from these rivers for their own consumption. In Tanzania, about 15% of the estimated hydro potential of 4,700 MW has been exploited. Rivers Rufiji, Mara and Kagera and their tributaries present Tanzania with a range of hydro potentials suitable for both small and large systems. On river Rufiji alone, hydro potential of over 2000 MW of installed capacity has been identified. The capacity exists for expansions of hydro projects at Rumakali and other existing hydro stations.

From Table 3.3, it is clear that, compared to Uganda and Tanzania, Kenya does not have a high potential for large hydo schemes but has significant potential for small and micro hydo schemes. Kenya would therefore be better placed to develop small and micro hydropower facilities on its numerous small but permanent rivers scattered all over the country particularly in the western part of the country, which at present does not have hydropower plant. When completed, the 60 MW Sondu Miriu hydro station will be the only large generation facility in western Kenya. Other hydropower potentials on river Tana are not included in Table 3.3 but are possible extensions of the existing Seven

Table 3.3. Hydropower Potentials in Kenya and Tanzania

Kenya	MW	Tanzania	MW
Sondu Miriu	60	Upper Kihansi	120
Ewaso Ngiro	90	Rumakali	222
Ewaso Basin	90	Ruhidji	358
		Masigira	118
		Stiegler's Gorge I	300
		Stiegler's Gorge II	750
		Stiegler's Gorge III	350
		Mandera	21
		Mpanga	160
Total	240	Total	2,399

Forks system, a move that may not be advisable given the vulnerability of the facilities to drought and the dire consequences as has been experienced in the past.

In general, East Africa has a vast hydro potential in excess of 6,000 MW that can supply all electricity requirements in the region at the present total peak demand of about 2,500 MW. However, achieving this through single national grid power network using large hydro stations would not only be too expensive for the economies of the regional states but would also cause far reaching impact on the environment. Maintaining the current networks with very small national coverage is already proving to be difficult. Small hydro potential existing in the region, if developed as isolated units or grid connected when economically beneficial, offers good chances for rural electrification. It is worth noting that of the total World's hydropower potential, 27% is in Africa (about 780,000MW) while South America has about 20% (577,000MW). The remaining 53% is shared among the rest of the world (Western Europe, United States of America, Canada, Eastern Europe, Asia etc). However, Africa has harnessed only less than 5% of its hydro potential while South America, with comparatively less potential, is using more hydropower than Africa. The African continent needs to learn from the experiences of countries like China, which has a vast territory, served by several small and medium hydro systems, some of which are operating in isolation. So far more than 40% of the Chinese hydro potential has been developed and there are about 50,000 installed small hydro units producing a total of about 6,000 MW mainly for rural electrification. The unique Chinese approach to rural electrification is the use of Micro hydro schemes with capacities of up to 100kW (0.1MW).

In fact, there are several factories producing even smaller systems of up to 5kW, referred to as Pico hydro schemes, which are exported to other countries especially to the Far East for rural electrification. Such machines are simple to install and require less than 5m pressure head for their operation and weigh less than 50kg. The cost comes to about USD 300 per kilowatt when installed. Viet Nam has been one of the major consumers of the Chinese so-called family hydro systems of 50W to 1000W capacity. Today similar systems are manufactured in Hanoi and are readily available in the markets. These simple small machines require only up to 2.5m of head and flow rate of just about $0.02m^3s^{-1}$. The 1000W system, for example, requires 2 to 4m head and flow rate of about $0.08m^3s^{-1}$. The technologies for these Micro and Pico hydro schemes are well developed and, if adopted by the developing nations in Africa, can make significant contribution in rural electrification. East Africa is capable of developing its vast small hydro potential to power practically all rural trading centers, schools, provincial and regional administration facilities, all fish landing beaches around lake Victoria and any other center of interest. Most of the already developed hydro schemes in East Africa are in 100MW range. In general, hydro facilities are state controlled and supply about 60% of the total electricity consumed in the region.

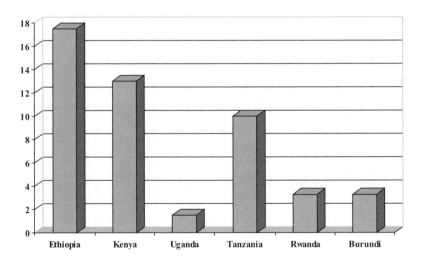

Fig. 3.3. Installed Mini Hydro Schemes in East Africa and some neighbouring countries

It has been shown that private sector has, in recent years, been actively importing and implementing schemes based on solar PV, diesel and petrol generators and this demonstrates a suppressed demand for electricity by the state controlled utilities. The suppression is policy driven and the requirement that consumers pay a standing charge on grid electricity supply has been a key factor in discouraging potential electricity consumers. The high connection fee paid by consumers without acquiring the right to own any of the components is another way of suppressing the demand. State owned utilities are not likely to voluntarily change these policies but mini hydro schemes with user friendly payment packages and operated by communities and independent electricity providers can definitely make a difference for the rural communities in this region. Figure 3.3 shows installed mini hydro facilities in East Africa and some neighbouring countries. These are far below the available capacities.

3.3 Co-Generation

Co-generation is one of the methods of generating electricity that have the best chance of success in the region particularly in rural electrification efforts. It is already practiced by large agro-based industries such as paper, sugar and timber industries but power generated has been exclusively used by the same industries because there is no arrangement for the sale of this energy to the public (see Fig. 3.4). These industries use bulky agricultural products as their principal raw materials and therefore, to encourage farmers to grow these inputs, transport cost must be kept as low as possible so that both the industry and the farmers can mutually benefit. For this reason, the industries are normally located in the rural areas close to the sources of raw materials. Their potential to generate electricity as independent power producers can be further developed with little additional investment since most of them are using the bye-products of the raw material to produce small quantities of electricity. However, the power generated cannot be sold to a third party because the operating licenses and the electricity distribution laws and regulations did not, in the past, allow such establishments to operate as power producers. Another important aspect is that the government can plan and choose the right location for the establishment of such factories so that rural electrification plans are also addressed besides the core function of the companies. The Companies

Fig. 3.4. Kisumu Molasses plant with un-tapped co-generation potential

also benefit by selling both energy and the core product instead of just one commodity. Thus the Companies would afford to lower the prices of both power and the core commodity and still make enough profit. The potential for co-generation is high and the prospects are very good for the Companies. East Africa has a number of agro-based industries that could supply electricity to their neighborhoods if suitable power legislation and regulatory arrangements are put in place. The area that could benefit most from the co-generation is the lake Victoria region of East Africa where there is a high concentration of agro-based industries, which have been producing electricity for their own use. There are about eight large sugar factories and at least five allied industries (molasses and paper industries) around lake Victoria whose energy outputs could make a significant contribution in the development of the region. Some of these factories have energy production potentials of up to 30 MW. The potential for electricity generation from bagasse within the Kenyan side of lake Victoria basin is estimated to be about 300MW but could be more than this if the factories are upgraded to higher production capacities. The exploitation of such capacity in the region would significantly improve the diversity of national power supply and save foreign exchange currently used to import fossil fuels used in power generation. Table 3.4 gives some examples of co-generation capacities in Tanzania. Most of the companies can upgrade their capacities and are in deed planning to do so in order to take advantage of the new energy policies. For example, Tanganyika Planting Company (TPC Ltd), which for many years had an installed capacity of only 4 MW recently (2005) up-graded its capacity to 20 MW and is planning to sell the excess power to the national grid.

Table 3.4. Examples of co-generation capacity in Tanzania

Power Plant/Company	Capacity (MW)
Kagera Sugar Company	5
Mtibwa Sugar Estate	4
Kilombero Sugar Company	3.4
Tanganyika Planting Company	20
Sao Hill Saw Mill	1
Tanganyika Wattle Company	2.8 (already selling power to TANESCO)
Total Capacity	36.2

The capacities indicated in Table 3.4 are based on the energy requirements for the Companies but they have greater potential than these and can be up-graded to much higher capacities within the present operational limitations. Other factories with co-generation potentials are Sony, Muhoroni, Chemelil, Miwani, Nzoia and Mumias Sugar Companies (see, e.g., Fig. 3.5), all located in western Kenya within lake Victoria basin where the only operational hydro facility is the 2 MW Gogo power station. Kakira Sugar Company in Uganda is also close to the lake and also to Uganda's main hydro power station at Owen Falls. Its power output can cost-effectively be connected to the Ugandan national grid network and, in deed, it has plans to sell the surplus of its 15MW production capacity to the national grid. In Kenya, Mumias Sugar Company in Western Province is set to be the first agro-based industry to venture into full scale commercial co-generation as it prepares to start selling its power to the distribution company, KPLC under the newly introduced provisions for independent power producers. Co-generation therefore has a great potential in the region particularly the central part of East Africa (lake Victoria basin), which has a significant co-generation potential. In general, however, the distribution of co-generating facilities in the whole region and possible choices of the locations of the new ones make them appropriate centers for rural electrification programmes.

Fig. 3.5. A sugar factory in Lake Victoria region

3.4 Portable Generators

Large fuel fired generators are the most attractive electricity generation facilities for independent power producers even though they are more expensive to operate since they require specific fuel inputs. About 90% of the population of East Africa has no access to grid power and there are arguments that even if they had, it is unlikely that they would use it. This is partly because of the high level of poverty amongst the rural communities that force them to live under conditions that would be even more dangerous with electricity supply and partly due to policies that suppress demand. So the issue of electricity is not a simple problem of lack of access and high cost but also a social issue as well. The use of thermal generators therefore will not help in bringing the cost of electricity to affordable level for the rural population. However, the prospect of ownership and personal control of application and running cost is an important feature for rural farmers who periodically receive large incomes and can afford to buy portable generators. So the potential for small portable fuel-operated generators that can supply enough electricity to a household is definitely available in the region. A number of these are already in use despite their high operational cost and purchase price. They are usually used only when necessary, for example, to light a wedding, funeral or an important social function. Their widespread use will depend on how low their prices will go but currently this is the option taken by the few medium and high-income groups in the rural areas. But even for these groups the use of generators is limited to special occasions due to high running costs.

3.5 Natural Gas, Coal, and Geothermal

Geothermal, coal and natural gas resources are not well distributed in the region: Uganda has none of them while Kenya, with some geothermal resources, has neither natural gas nor coal and Tanzania has no geothermal but has some deposits of coal and natural gas. The search for more of these resources is going on in all the three states. The available natural gas in Tanzania is estimated to be about 30 billion cubic meters at Songo Songo Island and at Mnazi Bay. The gas can be used to generate a significant amount of electricity in Tanzania and plans are already underway to modify existing thermal generators to use it. Some studies have however indicated that operating natural gas-fired generators are too expensive and would raise the cost of electricity. Deposit of Coal at Mchuchuma in South West Tanzania near the northern tip of Lake Nyasa has the potential to generate 400 MW of electricity for up to 40 years. The Kiwira coalmine has additional potential to generate more power. So far both gas and coal have not made any significant impact in the national energy scene especially at household level but they have a great potential in electricity generation for the national grid. Given the vast territory of Tanzania and the scattered nature of its towns and villages, these resources hold the future for rural electrification aspirations. Although some studies have indicated the possibility of finding geothermal sites in Tanzania, there has not been any concrete exploration plans. Similar studies have been done in Uganda with strong indications of geothermal potential estimated at about 450 MW in Buranga, Katwe and Kibiro areas but these have not been exploited. Kenya is the only country in the region that is already harnessing geothermal energy and is using it to generate over 100 MW of electricity from Olkaria geothermal stations in the central part of the Great Rift Valley. Kenya Electricity Generation Company (KENGEN) in collaboration with other independent power producers continue to search for more geothermal sites. A number of exploration wells have been dug and there are good prospects for finding exploitable sites. Such studies have been done around the present Olkaria area and also in Eburru field. Studies have also been carried out around lake Baringo as a result of natural occurrence of hot springs in the area but there have not been any conclusive findings. According to the present indications, geothermal potential in Kenya is in excess of 1000 MW and there are plans to harness another 600 MW by 2017. It is expected that the increased generation of electricity from geothermal sources will reduce the unit cost of electricity in Kenya because geothermal generation is considered to be the least cost energy in the region. Should more coal and natural gas be discovered, it is most likely that they will be used to generate electricity and not as domestic energy sources. The trend for coal and natural gas applications in Tanzania, a country with scarce biomass resources, is a good indicator for this assumption.

3.6 Solar Energy

The technology of solar energy conversion to electricity was virtually unknown in East Africa before the world oil crisis of the early 1970s, which triggered the need to search for other alternative energies. With a large percentage of urban population and almost all rural population having no access to national grid electricity, solar energy could play a significant role in domestic energy supply particularly for lighting. In the early phase of growth of the PV market in East Africa, the majority of the components for the systems were imported with the help of foreign donor funds. During the 1980's a domestic manufacturing expertise was gradually developed for solar thermal conversion while at the same time research on photovoltaic materials and production was encouraged in local universities. Although in East Africa not much was achieved in these areas, principally due to lack of institutional support, worldwide technological improvements however contributed to steady cost reduction of PV components. Potentially, a very large market for PV systems exists in Kenya, Uganda and Tanzania but to date implementation and their applications have been confined to affluent sections of society because of inadequate information and lack of distribution points in the rural areas. In 1996, it was reported that in Kenya about 40 000 - 60 000 households had installed solar energy systems, comprising more than 1 MWp of PV power. In addition to such domestic installations, over the past ten years, several hundreds of PV refrigerators have been installed for safe storage of vaccines, several water-pumping projects have been initiated and programmes to make low-cost solar devices such as solar lanterns have been initiated in the region. The total population in the three East African countries has gone past the 90 million mark and is steadily increasing. The majority of these people need electricity mainly for lighting their homes and solar energy seems to have the solution for them. It is however important to develop local capacity for installation, maintenance and after sales services in order to build user confidence in solar energy devices (PV systems, solar cookers, solar water heaters, solar dryers and solar lanterns etc). Sunlight is abundant in the region and with proper promotional activities, easy purchase terms and good incentives from the governments, solar energy could hold the answer to rural electrification programme. Independent power producers could also be attracted to photovoltaic electricity generation. It is estimated that at the rate of 100 W photovoltaic panel for every 4 people, there would be 2400 MW of solar electricity to light every home in the region and this would cost about 7.2 billion dollars at the present price of $3 per watt. Kenya is already generating about half of this but serves less than 10% of her total population. Obviously it has invested billions of dollars in the exiting generating facilities. According to 2003 estimates, of the 140,000 square metres of solar heat collectors for domestic water heating in Kenya, only 10% is for household use, the rest are installed in hotels, hospitals and institutions such as schools and colleges. The potential for solar power is indeed enormous but unfortunately this source is not given prominence in all the energy development plans of the three East African States. The fact that solar energy devices give the user the opportunity to control their use and the satisfaction of ownership would make them attractive to the rural conditions if people are sensitised and given the correct information on their use and maintenance. With enhanced state support, it is expected that the rate of solar energy application will considerably increase. The average daily solar radiation in the region is about $21MJm^{-2}day^{-1}$ with a minimum of $15MJm^{-2}day^{-1}$ and a maximum of about $25MJm^{-2}day^{-1}$. Given these conditions and a large population of more than 80 million people without electricity, the potential for solar photovoltaic home systems is virtually untapped.

3.7 Wind Energy

We have seen that there are some wind generators that are working quite well in the region. This is a clear indication that wind power can be successfully harnessed in East Africa. The highlands, lake Victoria basin and hilly regions, all have average wind speeds that are suitable for both power generation and water pumping. The few wind generators and wind pumps that have been installed have operated satisfactorily, producing power at reasonable levels. This is clear evidence that there is reasonable potential for wind energy applications in East Africa. In Kenya, two local manufacturers, one near Nairobi and the other in Coast Province have installed most of the over 350 pumps in the country. There are also a significant number of locally produced wind pumps in Tanzania. Using wind power to generate electricity has not been as widely practiced in the region as for water pumping. However, the two wind generators (150 and 200 kW) at Ngong near Nairobi (see, e.g., Fig. 3.6) and the 200 kW machine in Marsabit in Kenya as well as the 400 kW machine at Chunya Catholic Mission in Tanzania have all confirmed the viability of wind farms in the region. A small NGO known as Tanzania Traditional Energy Development and Environment Organization (TaTEDO), in its effort to promote wind generators, has installed a small wind turbine of capacity 600W at its centre with support from an external donor agency. In a recent development, the Tanzanian government stepped up support for feasibility study in northern Tanzania to establish whether a total of 50MW wind generators could be installed. The Danish organization that participated in the study confirmed that there is enough wind speeds in Mkumbara area to operate wind generators. Another study conducted at Setchet site by researchers at the University of Dar-es-Salaam also found the annual average wind speed to be 8.3 ms^{-1}, which is sufficient to generate electricity in the area. Attempts have also been made to assess the viability of wind generators at Mkumbula, Karatu where wind speeds were found to be, on average, $4.5ms^{-1}$. Thus there is enough evidence that there are good prospects for harnessing wind energy in Tanzania. Already reasonable effort has been made to use it for water pumping. For example in Musoma, Magu and Tarime areas, wind pumps are used for irrigation. There are also

Fig. 3.6. Wind generator operating in Kenya (Ngong Hill)

some wind water pumping machines in Singida, Dodoma and other regions of Tanzania. Wind energy options should therefore be more seriously considered since its use for electricity generation is likely to play an important role in rural electrification because it is relatively cheaper than oil-fired generation facilities, particularly in remote inaccessible rural areas. New begins

Figure 3.7 shows mean wind speeds for some selected sites in Kenya and it is evidently clear that mean wind speeds in the coastal region of East Africa are, in general, more than 4 ms^{-1}. The two sites at Malindi and Lamu are both in the coastal region and are over 100 kilometers apart but their mean wind speeds are almost the same. A similar trend has been observed along the Tanzanian coast. There are also a number of inland sites scattered all over East Africa that have mean wind speeds above 4ms^{-1}. Marsabit in Kenya, for example, has mean wind speed of about 11.5 ms^{-1} and there are other sites with similar high wind speeds. Thus there is significant wind energy potential in the region that should be harnessed.

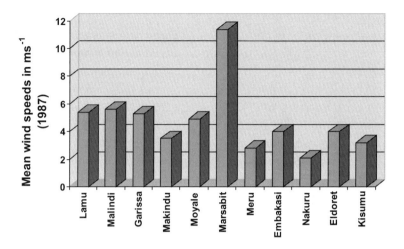

Fig. 3.7. Mean wind speeds in some sites in Kenya taken at ten meters above the ground

3.8 Biomass

When talking about biomass, we refer to energy obtained from direct combustion of any form of vegetation (wood, grass, shrubs, agro-waste, sawdust, sugar cane bagasse, etc) including their gaseous, liquid and solid products like methane (biogas), bio-diesel, producer gas, ethanol (power alcohol), cow dung and charcoal. Some electricity generators, the so-called thermal generators, use heat from biomass to produce steam that rotates the turbines. The co-generation discussed in the previous sections use heat obtained directly from biomass combustion. Some biomass materials such as sugarcane can be converted through fermentation into ethanol (also known as ethyl or power alcohol), which in turn can be used to blend oil-based fuels in order to reduce the consumption of oil particularly in the transport sub-sector. This method can help in reducing the amount of convertible currencies that poor developing countries are spending on importation of oil. Furthermore ethanol can be used in a variety of industrial productions such as chemical, pharmaceutical and beverage industries. East Africa is often overburdened by expenditures on oil imports when oil prices suddenly go up due to political turmoil in the oil producing countries. For this reason, the possibilities of producing ethanol from raw sugarcane or molasses and mixing it with oil for use in conventional petroleum powered machines and transport vehicles have been studied and prototype production schemes carried out. Biomass is therefore the most interesting energy source especially in the developing countries where both the government and most people are too poor to afford other sources of energy, which may not even be locally available. On a smaller scale, biomass can be converted into combustible gas known as biogas (methane) that is used for both cooking and lighting. In East Africa, biomass, in its various uses, accounts on average for more than 80% of the total energy consumption and almost the entire rural population in the region depends on biomass energy for energy supply mainly in the form of dry wood, also known as fire wood. One most important aspect of biomass materials is that, practically, most countries of the world can produce it in large quantities and this is part of the reason why it is such an important source of energy for rural communities in the developing countries; they have access to it and it is possible to generate it in any desirable quantities. In addition to energy, biomass is also used for many other purposes such construction of buildings and furniture and therefore it is generally in high demand. Fortunately these other applications demand high quality timber while for energy applications low quality wood such as dry branches of trees and other wood wastes are quite suitable. This means that wood fuel can be obtained from live trees without having to cut down the whole tree. Consequently the various uses are usually not viewed as competitive. The second interesting characteristic of biomass as an energy source is that it does not require special and complicated cooking devices (stoves) and traditionally the users, who are mostly women, design and construct the cooking place. This may be a simple arrangement of three stones between which wood is burnt or suitably dug out shallow oval hole in which biomass is burnt. The cooking pot is either supported by the three stones or is placed above the hole. More recently a number of portable improved wood stoves were developed and disseminated amongst the East African rural communities including schools and colleges. A fast-growing consumption technique is the conversion of woody biomass materials into charcoal, which is easy to store and transport. It is also consumed in smaller quantities than wood and is therefore most preferred by those who do not have space to store wood especially the urban middle and low-income groups. These groups together with the rural communities constitute more than 90% of the total population of East Africa. A large potential exists in the generation of biomass materials and also in the development of biomass energy devices like cooking stoves that can use various forms of biomass such as briquettes of different biomass materials such as sawdust, charcoal dust, fallen dry leaves, rice and coffee husks. Considering the number of people that use biomass as a source of energy and the important role played by biomass in producing other forms of energy, we see that biomass is such an important source of energy that should not be ignored in any rural energy planning process. Given that the situation regarding the present types of rural shelters, the level of poverty and the eating habits are not likely to change in the near future, biomass development must be seriously considered as an integral part of the energy systems. Its most important feature is the fact that it can be converted to commonly used forms of energy (heat and electricity) and also to gaseous and liquid energy sources such as biogas, producer gas, and ethanol. The level of these conversion technologies may be low in East Africa at present but there is great potential for their development. Production and use of charcoal, which is a high heat content biomass fuel, is well established in the region but the people need to be trained to develop and use efficient wood-to-charcoal conversion technologies. The general climate is also quite suitable for biomass regeneration programmes. Production of ethanol from biomass has been considered in Kenya as one possible way of reducing oil imports. This possibility was seriously considered in the 1970s when the oil crisis adversely affected the economies of East African states. At that time Kenya, for example, was spending as much as 35% of her total import bill on importation of oil. The idea that ethanol, produced from sugar cane, could be added to the imported oil and used to power existing vehicles without any engine modification gave Kenya some hope of reducing her expenditure on petroleum. A reasonable ratio of petroleum-ethanol mixture was expected to reduce the amount of imported oil by about 20%. Two alternative production methods were studied: one was to produce ethanol from molasses generated by the existing sugar mills in western Kenya and the other was to produce ethanol directly from sugar cane. The former was considered more economically viable since there were active sugar milling factories that could supply molasses to the ethanol plant. Two ethanol plants were therefore established within the western Kenya sugar belt: the Agro-Chemical and Food Company in Muhoroni and the Kisumu Molasses Plant. Apart from ethanol, these plants were also designed to produce other products of commercial value from the same molasses. East Africa has several active sugar milling factories that can adequately support ethanol production. However, the economic analyzes on local ethanol production showed that using the technology adopted by Agrochemical and Food Company at Muhoroni in western Kenya is more costly than imported gasoline. Thus ACFC can only make profit if ethanol is produced for export. But there are possibilities of reducing the cost of production to an economically viable level. Improvement of production efficiency can be made in areas such as fermentation, distillation, use of self-generated energy and acquisition of cheap or free molasses. This together with more cost-effective management arrangement would definitely make ethanol cheaper than imported gasoline. Therefore combining ethanol production with both electricity generation and sugar milling is necessary in order to substantially bring the cost of ethanol down. This is an option that is feasible in East Africa and should be given some in-depth consideration since there are a number of sugar milling companies that have the capacity to commercially co-generate electricity for public consumption. Thus the potential for biomass embraces a wide range of processing options from solid to gaseous fuels. The region however has no experience with bio-diesel although the potential for its development from locally grown crops exists. Research and development in this area should be considered and supported as a long-term energy plan for the region.

3.9 Concluding Remarks

Having identified the prominent energy resources in the previous chapter and noting that most of the locally available resources have not been developed, Chapter Three has attempted to expose the potential of various energy resources. It is evidently clear that only a small quantity of locally available energy resources have been developed in the region. There is still enormous quantities of various energy resources, particularly renewable energy resources, that are yet to be exploited. The main problems are mainly associated with the low level of local technological capacity in renewable energy technologies and the apparent complacency and hence inadequate support for energy development initiatives. Furthermore, information on specific energy sites including their commercial viability that could guide potential investors is not readily available. Scanty and sometimes inconsistent information is obtainable from [1, 3, 8, 9, 11, 17, 20, 22, 31, 32, 34, 43, 44, 60].

4

Energy Planning and Provision Challenges

4.1 Introductory Remarks

Since the three East African countries got their independence from the British colonial rulers in the early part of the 1960s, provision of energy has been one of the most sensitive and sometimes thorny issues in the region. The three governments have always tended to put too much emphasis on oil and electricity, both of which are playing a very important role in the transport and industrial sectors. However, at the household level, they have not made any significant impact since most households rely on biomass energy sources (mainly wood and charcoal). The few households, which are connected to the national grid, use electricity for lighting and operating household appliances but usually not for cooking. Very few people would use it for cooking. Most domestic needs for electricity do not require high voltage electricity. The large hydro power stations, which have been developed in the region, produce high voltage electricity and therefore target industrial applications. The high voltage is also good for long distance transmission due to power loses in the process. These characteristics make centralized grid electricity too expensive for household end-users, who generally are only interested in lighting and running of household appliances. A cheap decentralized generation facility that would target rural lighting and also stimulate small-scale rural enterprises that do not require high voltage would be more appropriate for rural conditions. The governments should not spend billions of dollars to extend national grid electricity lines to scattered rural settlements where the greatest need is for lights and home entertainments. It is therefore surprising that the governments have ignored these options. Perhaps this is why the governments found it necessary to impose state-managed monopoly on production and distribution of the two major commercial sources of energy - electricity and oil. What reason would a state give for ignoring the plight of its poor citizens? This and many more questions may be asked but the fact remains that the state will always be interested in the ventures that have the capacity to either pay large taxes or to directly generate huge revenues and there are good reasons for this. Another important aspect to consider is that despite the well-known fact that state businesses are generally mismanaged, leaders still found it attractive to allow the state to manage certain businesses in total disregard to the large losses incurred. In recent years, the states of East Africa privatized a number of state institutions that were essentially doing business rather than providing suitable environment for private enterprises. The energy sector incidentally was not privatized until the so-called development partners imposed very tough conditions. It appeared that some of these utility corporations were occasionally used as sources of funds for politically motivated projects and also for quick accumulation of personal wealth. It is has become apparent in Kenya that the state monopoly, Kenya Power and Lighting Company, provided funds for questionable purposes, which were deliberately concealed as general losses made by the Company. Obviously the taxpayers' money would be used to offset such 'losses' and the accounting officer would be fully backed by the political leadership for the "good work" done. However, the problems of the electricity sub-sector have been more profound than just official stealing of funds. The major problems are associated with the overall management arrangement, regulatory framework and their relationships with the policy maker. In Tanzania, for example, TANESCO was the sole public utility responsible for electricity generation, transmission and distribution and was, at the same time, the policy maker and the regulator of electricity industry. This situation was quite similar to KPLC in Kenya and UEB in Uganda. But this was not the only problem; there was widespread dissatisfaction with operational performance and also the government's inability to finance further development of the sub-sector. It became necessary to change this trend by introducing reforms that would not only revitalize investment in the sub-sector but also increase electricity production, transmission, distribution and sales. Thus reforms in the power sector in all the three states were introduced, first, to correct anomaly in the institutional arrangements and secondly to attract private investment and increase efficiency in the electricity sub-sector. The costs of reforms, understandably, were to be met by the same development partners who previously supported the governments that dilapidated the sub-sector. To implement the reforms, the governments were expected to enact some laws that would pave way for the creation of other bodies to take up some of the responsibilities of the sole state monopoly companies and also allow the involvement of independent power producers in the generation of electricity. Some glaring questions that should be asked are: Will these moves increase accessibility to electricity in the rural areas? What planning considerations will be given to presently used non-commercial energy sources on which rural communities heavily depend? One of the strategic objectives of the energy sector is to ensure reliable, accessible and affordable energy and therefore diversification of energy sources becomes important so as to ensure that natural disasters like drought have very little effect on overall production and distribution of electricity. Diversification of sources and attraction of private sector into electricity generation is expected to encourage the development and use of renewable energies such as solar and wind but the governments will have to make deliberate efforts to provide enabling environment for these to happen. Then there is the question of energy conservation for purposes of protecting the environment and resources. Economically recoverable oil has not been found in East Africa and so the region depends on oil imports mainly from the Middle East and the governments will continue to spend their limited foreign earnings on such importation. There is need to control this not just through pricing but also through efficient use of oil-based fuels and use of alternative sources. The transport sector is therefore an area that requires special attention with a view to reducing oil consumption. The implication of this is that roads and traffic rules must be designed to reduce fuel wastage in addition to diversifying modes of public transport and making them more convenient, comfortable and attractive in economic terms. These are far reaching challenges that mere reforms that are restricted to electricity generation and distribution cannot address. Furthermore, the reforms are not addressing the concerns of the majority of the population who simply require heat energy for cooking and a little kerosene for lighting and who have independently managed their own source of heat for many years. Kerosene is used by an increasingly large number of rural people because it can be bought in small quantities and the user can regulate its use. Access to electricity should also be arranged in such a way that the user pays for what is consumed only and not for all sorts of gadgets including meters which belong to the utility company. The level and payment of electricity connection fee should also be user friendly, for example, 50% of this could be distributed in the electricity bill for a period of one or two years. Introduction of prepaid meters where the consumer's supply is based on ability to recharge the meter could also encourage people to apply for grid connection. This method was introduced by TANESCO in Tanzania and there were signs of increased demand even among the low-income groups.

Energy provision challenges in East Africa can be seen in the following contexts:

- **Legislative aspects, which are very unfriendly to and exploitative of the consumer**: For example the imposition of the mandatory minimum standing charge that the user must pay even if there is no consumption of electricity during that period. This has been a source of many complaints and has discouraged many prospective clients from getting connected especially in Kenya. The second example is the requirement that in order to be connected, the applicant must pay the cost of all the materials needed for the extension of power to the consumer's premises.

- **Legislative aspects, which discourage investors from participating in the electricity sub-sector**: Although this is expected to be corrected through proposed reforms, it is unlikely that the new rules will allow independent electricity transmission and distribution companies to operate. Reforms that are restricted to the generation and sale to a single utility monopoly will not have the impulse that is required to liberalize the electricity sub-sector.

- **The level of poverty of the people**: The level of poverty is relatively high in the region especially in Kenya where more than 50% of the population lives below poverty line. In both Uganda and Tanzania about 36% of the total population live below poverty level. This is a major drawback in the effort to improve electricity accessibility in the rural areas.

- **The land ownership laws that encourage the establishment of scattered settlements or establishment of "homes"**. This has always made it too expensive to implement electricity line extensions to the consumers so that even those who can afford to pay for it are normally reluctant to get connected.

- **Electricity pricing policy**: there are many factors that are considered in pricing electricity and practically all the costs are passed on to the consumer. Most of these costs are based on long-run marginal costs including fixed operational and maintenance costs and variable consumable such as fuel margins and overhead costs. In addition to these there are taxes and levies, which may be imposed to carter for further developments such as rural electrification. The disappointing fact is that although consumers pay for all these expenses and levies, the money is hardly used for the intended purposes. The actual tariffs, however, vary from country to country. For example, Tanzania implemented tariff reductions especially for industrial customers but this only brought them down to the level of the Kenyan tariffs. In Uganda the Tariffs are much lower and in this regard many new industrial establishments in the region prefer to set up their businesses in Uganda.

- **Over reliance on hydro as the main source of electricity**: This has forced utility companies to occasionally ration electricity due to low levels of water in the reservoirs. Kenya has been the most affected by this because its five major hydro stations are on one river and, worse still, one dam is the main reservoir for all the five stations.

- **Frequent electricity interruptions**: Servicing and maintaining electricity grid network continue to be a serious problem for the utility companies. These interruptions often cause power surges that damage electrical appliances and it is estimated that consumers lose millions of their hard earned money through such power surges. In Kenya, an average of about 11,000 unexpected interruptions of electricity supply are reported every month but the actual number of cases is higher than this since many cases are not reported. Losses are also incurred through illegal connections which are estimated to cost KPLC over 15,000 US dollars per year.

- Finally, there are the perennial management problems of general inefficiency including rampant misappropriation of funds.

This last context appears to have been the main reason why international development partners applied pressure on governments to introduce reforms into the electricity sub-sector. The desire was to dismantle the electricity supply structure, which was dominated by vertically integrated state-owned electricity utilities, and open the sub-sector up for independent investors most of whom would come from the more developed countries. If the reasons were more than these then obviously all energy resources of significant values would have been considered and more emphasis would have been put on rural electrification. But the reforms do not address these directly although they may have some implications on the development of other energy sources and also on rural electrification. It has all along been evidently clear that the monopolistic management structure was a significant contributor to the under-performance of the region's power utilities that were characterized by unreliability of supply, low capacity production, deficiency in maintenance, high transmission and distribution losses, high cost of electricity and inability to mobilize sufficient investment capital. These were bad enough by any standards but the question is: Would the proposed reforms address all the above problems? And for how long and to what extent are the poor developing countries going to depend on oil which they have to import at increasingly heavy sacrifices? It must be noted that more than 75% of the world population is in the developing countries and any increased demand and use of oil in these countries will have dire consequences on the environment by accelerating climate change and global warming processes to unprecedented levels. Increased demand will also raise the price of oil and hence greater sacrifice for the developing countries to import oil. In addition, world oil reserve will rapidly decline if substantial new reserves are not discovered. Care therefore must be taken in handling the use of oil and other fossil-based fuels. The energy reforms in East Africa appear to have downplayed future implication of increased oil consumption on the economies of the three countries and also on the environment. As world oil reserves decline, it would not be surprising if the next energy reforms imposed limits on oil consumption.

While making efforts to address the challenges, due regard must be given to the protection of the environment, the expansion of energy infrastructure, firm security of supply, diversification of sources and accessibility to all sectors of the population with special attention to the rural population. All these should be considered as aspects of energy provision challenges. In addition to these, the people must be made to understand the need for energy conservation.

4.2 Reforms: The Immediate Challenge

Compared to other parts of the world, East African power sector reforms have been slow and to a large extent, limited to divestiture of the traditional state-owned power corporations and entry of independent power producers. The aim was to meet shortfalls in electricity generation and also improve efficiency of supply. These are issues that are of great interest to the manufacturing and general industrial activities but may mean almost nothing to the more than 80% of the population who have no access to electricity. One may argue that although these people have no access to electricity they are also beneficiaries of industrial output and other electricity-supported products. However, the global power supply concerns are about people and, naturally, large scale availability of power to the people also creates the desire to set up new industries. The Power sector reforms that have been carried out in East Africa basically targeted and indeed affected the electricity sub-sector. The major turning point in this process was the amendment of electricity acts in the second half of the 1990s by the three states of East Africa. It was these amendments that purportedly created the enabling environment for the operations of independent power producers and the unbundling of the state-owned monopoly companies. The individual national parliaments in 1997 and 1999 passed these amendments in Kenya and Uganda respectively. In Tanzania, the process started in 1992 when the government changed its policy to allow the participation of private sector in electricity generation. In 1999 a new electricity policy was introduced and in 2001 an Act of Parliament established Electricity and Water Utilities Regulatory Authority (EWURA). The impacts of these policy changes have been measured in terms of the number and capacity of independent power producers (IPPs) in relation to the traditional suppliers. Indications are that Tanzania and Uganda will give the independent power producers the major share of electricity generation under more favourable environment than Kenya and therefore fewer IPPs are showing interest in operating in Kenya.

It is clear from Fig. 4.1 that both Tanzania and Uganda have attracted a lot of interest from independent power producers. In Uganda there are IPPs interested in hydro generation of electricity while in Kenya there are some who are already involved in using geothermal as their source of electricity. Most of the independent generators, however, are engaged in fossil-based generation methods, which obviously will increase national expenditure on oil imports. In Uganda and Tanzania,

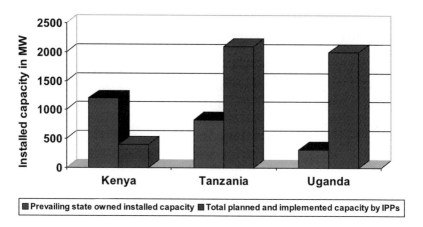

Fig. 4.1. National capacities compared with planned and implemented capacities by Independent Power Producers

the high hydropower potential and governments' policy on their development and management appear to be the main attraction for IPPs. The presence of coal and natural gas in Tanzania may also appeal to some IPPs but, in general, the institutional arrangement for managing the energy sector is an important factor that most private investors would want to consider. In terms of legislation and the restructuring of institutional arrangements in the management of the energy sector, the previously single control by KPLC in Kenya, TANESCO in Tanzania and UEB in Uganda have been removed by the creations of new bodies to deal with generation and regulatory matters. A brief look at the amendments of the electricity Acts indicates remarkable differences in the emphasis on various issues such as rural electrification, tariff changes and possible levies. For example, the Kenyan Act addresses the issue of access only to a limited extent and trivially mentions rural electrification process while giving the minister concerned powers to introduce levies and decide where and how rural electrification is to be conducted. The minister may establish rural electrification fund, but there is no firm guidelines on this. The Ugandan Act, on the other hand, puts more emphasis on electricity access in the rural areas and gives the Minister responsible the authority to develop a sustainable and coordinated Rural Electrification Plan and Strategy, establish Rural Electrification Fund and determine the criteria and appropriate level of subsidy. Furthermore, the Minister is expected to keep and maintain National Rural Electrification Data Base to assist in monitoring the progress. It is clear that while Uganda is showing a lot of concern for rural electrification, Kenya does not seem to be interested and does not even mention the possibility of subsidy. In Tanzania, the Rural Energy Master Plan includes renewable energies and proposals to set up Rural Energy Agency and Rural Energy Fund. These are expected to spur rural energy development but there are doubts that, with privatization of TANESCO, the distribution companies will have any incentive to carry out rural electrification. These reforms have been in place long enough to determine their impacts on electrification status in the region in order to assess the real challenge in this respect. In Tanzania, the reform process started in the early 1990s when the country was in transition from socialist economic policy to free market economy. Its case is therefore slightly different from that of Kenya and Uganda and may not provide the real impact of reforms that were proposed by the World Bank. Examples are therefore taken from Kenya and Uganda.

4.2.1 Effects of Electricity Reforms in Kenya

It is true, to some extent, that since the reforms were initiated the performance of the state corporations that managed the electricity sub-sector has significantly improved. Some of them have cleared their past huge deficits and are beginning to make profit. For the case of TANESCO this improvement could be explained by the fact that professional management consultants were hired to run it and therefore had to work hard to prove their competence. For both Kenya and Uganda, the improvements could be attributed to the governments' introduction of tough measures against inefficient Chief Executives of the companies. The Kenyan case is a clear case where a new party came to power and took drastic and stringent actions against Chief Executives who knowingly misappropriated public funds. This is a pointer to the fact that gains could be made and management improved without the introduction of the reforms. This is not to say that the reforms were not necessary. The reforms covered a wider spectrum of issues than just profitability of these companies. The global concern is to see that the more than two billion people who have no access to electricity are connected. One would therefore expect that since the introduction of electricity reforms in East Africa, there has been a significant change in the rate of electrification levels. This has not been the case in Kenya. It looks like despite the reforms, business has gone on as usual. Figure 4.2 shows this very clearly that the situation has not changed from what it was before the reform-motivated electricity Act amendment. Of course, instant changes should not be expected but some signs should be begin to appear three to four years later. The trend of electrification levels for all categories of people continued without any response to the policy changes. This reinforces the view that the reforms have essentially emphasized the development of IPPs and improving financial performance of state-owned utilities at the expense of rural electrification. In real terms the number of people without electricity continued to increase in both rural and urban areas. It is therefore clear that the new policies do not comprehensively address rural electrification strategies and there are no guidelines on how it should be implemented. In the long run, there may be tangible sectoral benefits of these reforms but this may not trickle down to the majority of the citizens unless reforms are accompanied with comprehensive measures to reduce poverty in the rural areas.

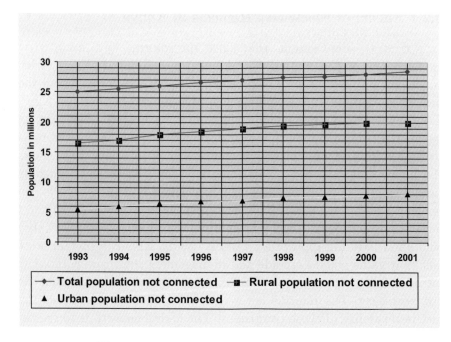

Fig. 4.2. Population without electricity in Kenya

Another measure that is used to determine the extent to which the reforms accelerate access to electricity are electrification rates, which refer to the number of new domestic connections in a specified year compared to the number in the previous year. These rates have remained very low in Kenya in both the periods before and after the policy turning points, indicating that there has not been any impact of the reforms. In fact, in certain cases, the figures were more impressive prior to the amendment of the Electricity Act. It should however be noted that there are other factors that affect the electrification rates such as the fluctuating cost per connection and increased population. It seems that the arrangement where KPLC funds Rural Electrification while at the same time free to use the same funds to offset any operational losses has made it almost impossible for rural electrification programme to succeed. The logical conclusion is that the management of Rural Electrification Programme in Kenya has been ineffective due to the government's lack of interest in its progress. This is confirmed by the fact that the proposal to establish rural electrification agency came at the end of the reform agenda and appeared to have been an afterthought given that it was not even provided for in the Electricity Act. Kenya Power and Lighting Company will continue to play a key role in rural electrification process and given that it was only recently salvaged from near collapse it is likely to pay more attention to the improvement of its operations rather than to rural electrification. To maintain some sort of growth it is likely that KPLC will continue to concentrate its distribution activities in urban areas where the extension of grid power line is cost effective due to high concentration of consumers in one area. Consequently the future of rural electrification in Kenya will continue to be very bleak.

4.2.2 Effects of Electricity Reforms in Uganda

Access to electricity in East Africa is lowest in Uganda with less than 5% of the total population electrified. As in the Kenyan case, consideration is given to the population not connected in urban and rural areas and also the overall situation in the country. Figure 4.3 gives details of these from 1996 to 2002 [22].

There has been a general trend of marginal increases in accessibility to electricity since 1996 within the population categories examined and this trend continued even after the amendment of Electricity Act in November 1999. Like Kenya, the changes in electricity management

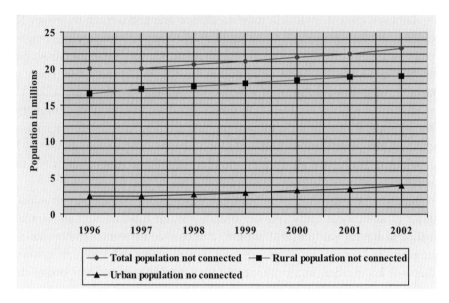

Fig. 4.3. Population without electricity in Uganda

policy as a result of the amendment of the Act have not made any difference in the situation of the majority of people living in the rural areas. Although there is a plan to implement the Energy for Rural Transformation Project in order to increase rural electrification levels, it is unlikely that this will significantly benefit the rural communities. This doubt arises from the fact that the target set for this project is too low at 10% by the year 2012 indicating lack of commitment and seriousness in addressing rural electrification. Evidence from other African countries shows that for the same period of time it is possible to achieve almost twice as much as this target. Population growth rate, high tariffs, rigid payment modes and the high poverty level will reduce the quoted target to a mere business as usual status and, given the present trend of household electrification rates, even this modest target will not be realized. It is important to note that rural electrification rates cannot be expected to significantly increase without addressing the socio-economic circumstances of the people. In Uganda, the reforms are at an advanced stage but it appears they were undertaken primarily to implement the privatization of the government controlled utility companies. Very little has been done in the area of rural electrification even though there is apparent incentive given by the new regulatory and policy frameworks. The arrangement for rural electrification that has been put in place is basically similar to what has existed before the reforms. It is only formulated in a slightly different style while the government control and involvement is still firmly in place. For example, Rural Electrification Board is headed by the Permanent Secretary in the Ministry responsible for energy and it is known that this arrangement limits the autonomy of the Board and mixes up policy, financial and implantation matters to the extent that there is no proper monitoring of the progress. This was the main cause of the failure of Kenya's rural electrification policy and Uganda is unlikely to succeed with the same approach despite the good statements of intention.

The two examples from Kenya and Uganda are indicative of the major challenges in the electricity sub-sector in East Africa and confirm that, despite the purported changes to improve the electrification levels, the so-called common citizens of this region will not benefit from these efforts. The truth is that the restructuring of the electricity sub-sector is deliberately designed to improve the financial performance of state-owned corporations, obtain addition revenue from IPPs and increase electricity security for the industrial and manufacturing sectors. To achieve some level of success in rural electrification, the governments must specifically encourage generation of power in the rural areas using resources that are available in those areas particularly small hydro, co-generation, biomass, solar and wind energies while at the same time seriously addressing rural poverty. Without these, rural electrification will remain an unfulfilled dream for many years, if not forever. At present, there does not seem to be any seriousness on the part of the governments to tackle the issue in an organized manner, as there are no proper records regarding rural electrification status. Even data on rural energy resources and their potentials are not adequate and the little available are not accurate enough to assist potential investors. Official energy targets are given as conditions for issuance of licenses but are never followed up thereafter.

In addition to the electricity issues, there are also challenges in the oil sub-sector to deal with. The region has no recoverable oil at present and yet there is a growing demand for kerosene in the rural areas particularly for lighting. It is fortunate for the governments that presently a large fraction of rural population is not economically empowered to use kerosene freely and therefore this puts a limit to the amount of kerosene used. A situation in which the demand for kerosene can double will put a great deal of strain on the economy of the region and so the governments must prepare alternative energies now and not later. The alternative energy sources that should be considered for the region are solar, biomass, decentralized small hydro and wind. The challenges of these sources lie in their technologies and that is the subject of the next sections.

4.3 Concluding Remarks

Proper energy planning in the region is evidently far from adequate. Various energy issues including some energy resources are still under government departments that are not concerned about energy problems. This is making it difficult to formulate clear energy policies since the departments have their priorities in different issues. In this regard, energy provision policies are not well coordinated and hence very difficult to implement. The fact that some energy resources are not commercially recognized at the national level creates an imbalance in the level of support given for the development of different energy resources. The tendency therefore is to unnecessarily put more emphasis on control and management of commercial energies that are of economic interest to the nation. The sad thing about this is that commercial energies such as oil and grid electricity benefit only a small fraction of the total population. More information regarding energy planning and provision problems including policy reforms are given by [6, 9, 10, 12, 22, 24, 28, 29, 32, 37, 39, 41, 42, 48, 49, 50, 52, 60, 61].

5

Alternative Energy Technology for East Africa

5.1 Introductory Remarks

We have seen that the very nature of socio-cultural practices and the rampant poverty in East Africa complicate the process of seeking solutions to the energy problems. To address the problem of power distribution, a carefully planned energy strategy must be implemented to take care of the scattered nature of rural settlements, which is basically a reflection of the land ownership policy and individual freedom of land-use choices. In addition to this, the poverty levels of the people must also be considered so that suitable power consumption tariffs and manageable payment terms are instituted. One obvious approach is to ensure that the people are given as much ownership of the power system as possible. This will inculcate some degree of participation and hence responsibility in managing the operations of the systems and this would enhance sustainability. Another similarly obvious step to take is to strive to produce energy within the community areas in order o avoid large expenses incurred in long distance transmission. These conditions point to the fact that the source of energy must have certain characteristics such as possibility of locating it at a suitable site with respect to the demand concentration centers. These requirements eliminate many traditional energy sources such as large hydro and geothermal which are location-specific, that is, you have it where it is available or you do not have it at all. A final consideration is the availability of crucial inputs within the local environment or the possibility of generating them locally. Given these characteristics, one cannot base long-term energy strategies on limited sources such as oil, coal and LPG especially if the country is not endowed with them. East Africa does not have oil and has only a limited amount of coal and natural gas in Tanzania. The only logical choices for the region are solar, wind, small hydro plants and biomass. However, efficient use of these sources require some technologies and corresponding critical mass of technical know-how that must be developed as part of the energy development strategy. Achieving these goals would not be too difficult for any serious government as all the four sources are in abundance in the region and to do this, the conversion technologies for these sources must be well understood. This will enable energy specialists and policy makers to identify areas for immediate attention and what steps can be taken to mobilize resources for their development. Attempts are therefore made in the following sections to give an account of the technologies for each of the four sources. Efforts have been made to simplify any complicated concepts without distorting the general meaning and special attention is given to solar energy because it is an abundant renewable energy in East Africa. Its development may make a remarkable difference in the rural energy situation particularly as lighting energy.

5.2 Solar Energy Technologies

The energies that ordinary rural person needs in order of priority are;

(1) high temperature heat,
(2) light,
(3) electricity basically for entertainment, and
(4) low temperature heat.

Top priority is given to high temperature heat for cooking and boiling drinking water while the second priority is given to lighting. The other needs are not as vital for survival as these two and so the big question is: Where can one readily get the fuel to provide high temperature heat and also fuel for lighting at affordable price? The ideal source would be one that satisfies both needs free of any charge. Since the application of any energy source requires the use of specially developed device such as stove and lighting gadget, the next question would be: which source is readily available at all times to warrant some investment on application device? The answers to these questions would determine the source of energy people would go for not just in the rural areas but also in the urban situations and this should be the beginning point for a serious energy supply plan. In East Africa, like in any other equatorial region, sunshine is abundant and is received in sufficient quantities on a daily basis. Globally, the amount of solar energy received on the earth's surface is about 1.73 x 10^{14} KW, which is about 10^{14} times more than the world's annual energy consumption. This means that if efficient conversion technologies can be developed to harness solar energy then the sun can supply all the energy needs of the world. For a long time the important role of the sun as an energy source was limited to communication satellites and general space exploration but the oil crisis of the 1970s brought the realization that it could also be an important source for both heat and electricity. However, since time immemorial, ordinary people have depended on the sun for drying agricultural products such as cereals, vegetables and fish in order to prolong their storage life. But this application was treated trivially as a natural process since simply exposing the product to direct sunlight in an open air without the need to use any special device could satisfactorily perform it. However, it is associated with product losses and contaminations due to wind, birds and animals. Furthermore, the heat is not efficiently utilized because large quantities of heat are lost by convection and radiation. If special devices are designed to efficiently convert solar energy into heat, then much higher temperatures can be obtained for a number of applications. There are also known technologies for converting solar energy directly into electricity, which can be used for lighting and operation of household appliances. We will discuss these technologies in the following sections.

5.2.1 Low Grade Solar Heat Applications

Low temperature heat has many applications even in the warm tropical climate. Some of these include drying of crops, fish and water heating. The economy of East Africa is based on agricultural activities and therefore food preservation is very important for individual families and even at the national level. Maize, which is produced in very large quantities is the source of staple food known as "ugali" in Kenya. Rice, another cereal crop, is the staple food for most communities living in the coastal regions of both Kenya and Tanzania while in most parts of Uganda, banana is the staple food. The preparation of practically all these staple foods require that the product is first dried and then ground into soft flour. The drying process needs low-grade heat, which can be and has always been obtained from sunlight without using expensive devices. In addition to the processing requirements, drying of these products down to very low moisture contents prolong their shelf life. Similarly, the popular lake Victoria sardine locally known as "daga" or "omena" is a highly perishable fish but when dried down to low moisture content, can be stored in good condition for as long as one year. Thus in the absence of electricity and cold storage facilities, solar drying becomes an important food preservation technique for the rural communities. Solar dryers have been developed for this purpose but unfortunately there has not been any effort on the part of the governments to support or even promote these technologies despite the fact that the principle of converting light into heat is very simple. Sunlight that is absorbed by any solid material is converted into heat inside the material. However, not all light falling on the material is absorbed; while some are absorbed, the rest are either reflected or transmitted. A transparent material will transmit most of the incident radiation and is therefore a bad absorber. On the other hand, a shiny surface will reflect a large fraction of the radiation and is also a bad absorber. For solar energy conversion into heat, we need a material, which is opaque and a poor reflector. Once this is achieved, the heat in the material must be transferred to the location where it is needed, e.g., into water, if the intention is to heat water, or into a room if space heating is the objective. This should be done without allowing too much heat to be lost in the process. We recall that as the temperature of a body is increased, without changing the temperature of its surrounding, heat losses through radiation and convection also increase. It is therefore important that these heat loss processes are suppressed as the solar energy absorber heats up. This means that the absorber material must be properly insulated while at the same time well exposed to receive solar radiation. Naturally the exposed face will be cooled by convection and radiation and so it must be covered by a material that would suppress these loses while allowing solar radiation to reach the absorber. Such a material is obviously glass or transparent plastic sheet with poor transmission of thermal radiation. Figure 5.1 shows a configuration of a simple solar energy thermal converter commonly known as a flat plate solar collector. The backside of the collector plate and the sides of the box of a more realistic collector would be thoroughly insulated to minimize heat losses by conduction. If it is for water heating then the plate would be connected to a loop of water pipes to carry away the heat from the plate. If, on the other hand, it is to be used for heating air for space heating or drying purposes then provision must be made to allow air to pass over, under or along both surfaces of the collector plate to pick up the heat and direct the heated air to where it is needed.

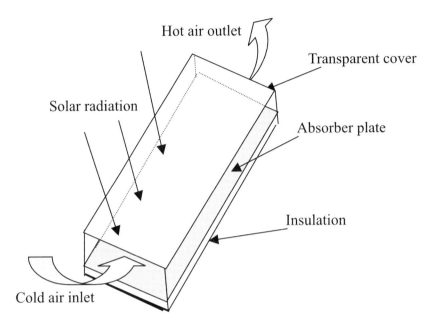

Hot air outlet

Transparent cover

Solar radiation

Absorber plate

Insulation

Cold air inlet

Fig. 5.1. A simple Flat plate Solar Collector

For air to move along the heater by natural convection, it has to be inclined to the horizontal so that the cold air inlet is lower that the hot air outlet. This will allow warm air, having been heated inside the box, to rise through the box by natural convection due to reduced density. As the warm air rises, more cold air is drawn in and the process continues. The sunlight that reaches the box is transmitted through the cover on to the dark collector plate which then becomes hot and in turn heats the air. In order to use this heater for drying, it must be connected either directly or by a duct to the chamber containing the product. The warm air draught through the system can be substantially improved if the dehydration chamber has a suitably designed chimney.

Figure 5.2 shows a complete solar dryer designed for dehydration of cereals and vegetables. When designing the shape of the dehydration chamber, consideration should be given to the shape of the commonly used grain storage structures in East Africa. This would make it easy for any local grain storage builders to construct and maintain the dryers. Farmers in some parts of East Africa use a number of these dryers. At a national level, the governments are using large oil-fired grain dryers to dry maize and other cereals for long-term storage. For example, in Kenya, the National Cereals and Produce Board (NCPB) use large oil-fired dryers at the grain storage facilities located in agriculturally high potential areas in the country. Cereals delivered to the national storage facilities are expected to be dry down to the set level of moisture content. If the product is wetter than the required moisture content, then the farmer is charged per moisture content above the set level. The product is then dried using centralized large oil-fired dryers. To avoid such expenditures, farmers could be encouraged to properly dehydrate their grains using these effective solar dryers before selling them to the NCPB. Alternatively, the Board could use large solar tent dryers whose operations are based on the principles of greenhouse effect. There are other designs of solar dryers but their modes of operation are basically similar to what has been described above and practically all of them strive to trap as much solar heat as possible inside both the heater and the dehydration chamber. The operation of the dryer will ultimately depend on how carefully the design factors are considered.

A lot of research has been done on various types of solar dryers and the results led to the development of different dryers. A number of these research activities were carried out in Kenya and Uganda but, apparently, the dryers proved to be too expensive for rural farmers. As is expected most of the activities were carried out without support from relevant government departments such as the Ministries in charge of agriculture despite the clear evidence that the dryers perform well and produce better results than the traditional open air sun drying besides reducing post harvest losses.

Another important low temperature solar energy application is for water heating. Like air heaters for drying purposes, solar water heaters also use the simple flat plate collector described above. Water heaters are normally fixed on top of the roof to get them out of the way, eliminate the necessity to attend to them and make sure that they receive maximum solar radiation everyday. This means that the water supply to the building must have high enough pressure to facilitate water circulation through the heater.

The temperatures in the highlands of East Africa can sometimes go as low as 14°C especially from early mornings to mid-mornings when most people are at home to use water. Water heating is therefore necessary in many parts of the region particularly in hospitals and ho-

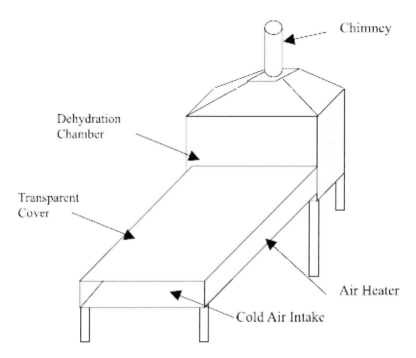

Fig. 5.2. Natural Convection Solar Dryer using a flat plate solar collector

tels. Solar water heating technique is slightly different from air heating method but the basic principles are the same. That is, the system must be able to trap as much solar heat as possible in the heater section of the device. The components of solar water heaters are the same as for air heaters but the design must consider the fact that heat is to be transferred from the collector o the water. Hot water should then be stored in a storage tank from where it can be drawn for use. Both the storage tank and the solar collector must be thoroughly insulated to prevent heat losses through conduction. A common design of solar water heater is shown in Fig. 5.3. Basically the water heater consists of transparent cover, flat plate collector under the cover and a hot water storage tank. The function of the transparent cover is to prevent convective cooling of the flat collector plate. Sunlight going through this cover is absorbed by the flat plate collector, which gradually gets heated up. The heat is then transferred to the water inside the pipes running along the collector plate. The design considerations must ensure that as much heat as possible is absorbed by the flat plate collector and then

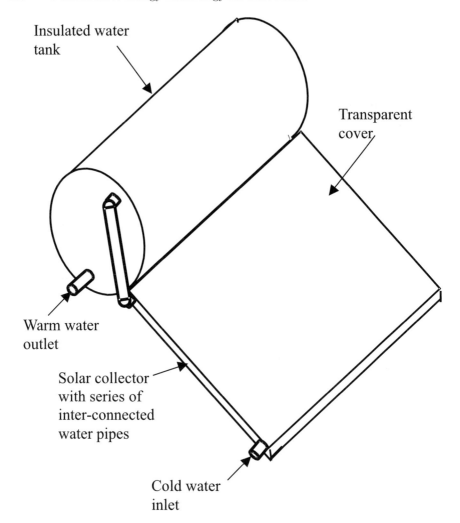

Fig. 5.3. A simple solar water heater with a flat plate collector

transferred to the water in the pipes. The reflectivity of the plate must therefore be reduced as much as possible. Basically there are two types of solar water heaters: an active one hat uses a pump to create enough pressure for water to flow through the system, and a passive one that works on the principles of natural convection flow where heated water, being lighter than cold water, rises through appropriate pipe into the storage tank. The passive one is usually referred to as Thermosyphon Solar Water Heater. The theories that determine these design factors are discussed below

The absorptance of the collector plate is a very important property for the performance of any solar heater and if possible the plate should have the following characteristics:

- Low reflectance of solar radiation.
- High absorptance of radiation with wave lengths in the range of solar spectrum.
- High emittance of the long wavelength infrared radiation.

A surface having these characteristics is known as a selective surface. The technology of producing selective surfaces is well developed and various types of selective surfaces are in use.

In order to design an efficient solar thermal converter, one needs to thoroughly understand what happens to solar radiation incident on the device by carrying out a heat balance analysis. Solar radiation absorbed by the plate should be equal to the useful heat obtained from the plate plus losses by conduction, radiation and convection. Heat balance analysis for the solar collector can be carried out using the properties of the transparent cover and the absorber plate. If energy that is reaching the transparent cover per unit area is I, then $I\tau_c$ is the transmitted and the amount absorbed by the plate is $I\tau_c\alpha_p$, where τ_c is the transmittance of the transparent cover, α_p is the absorbance of the collector plate and I is the incident solar radiation per unit area. Heat lost by convection from the plate to the air on the plate surface is

$$q_h = h(T_p - T_a) \tag{5.1}$$

Still in the same direction, heat is lost from the plate by radiation:

$$q_r = \frac{\sigma(T_p{}^4 - T_c{}^4)}{\dfrac{1}{\varepsilon_p} + \dfrac{1}{\varepsilon_c} - 1} \tag{5.2}$$

On the back side of the plate, heat is lost by conduction

$$q_k = \frac{k(T_p - T_i)}{x_i} \tag{5.3}$$

where x_i is the thickness of the insulation and T_i is its outside temperature. Finally, heat balance equation for useful heat obtained from the plate can now be written:

$$Q = I\tau_c\alpha_p - q_h - q_r - q_k \tag{5.4}$$

Giving the plate conversion efficiency per unit area as:

$$\eta_p = \frac{Q}{I\tau_c\alpha_p} \tag{5.5}$$

But the conversion efficiency for the whole solar collector is

$$\eta = \frac{Q}{I} \tag{5.6}$$

Using the flat plate collector as an air heater and inclining it to the horizontal as shown in Fig.5.2 enables the dryer system to create a suitable condition for natural air movement upwards into the dehydration chamber where it picks up moisture from the wet product. The cool moist air is then drawn by the effect of the chimney into the atmosphere. The effectiveness of the air heater depends, among other things, on the convective cooling effect caused by the conditions such as wind. Therefore when analyzing the performance of solar thermal converters, the determination of the convective heat transfer coefficient is one of the most crucial and delicate aspects. However for a well-designed heater this should not be a major concern.

5.2.2 High Temperature Solar Heat Applications

The low-grade solar heat applications use conversion devices that do not require any special optical arrangements to increase the intensity of sunlight. To obtain high temperatures from sunlight, there is need to provide a mechanism for increasing the intensity of solar radiation reaching the absorber material. This is normally achieved by using solar reflectors to concentrate radiation on to a smaller area. Thus, all high temperature solar devices have these concentrators, which are carefully designed to direct solar radiation to the desired area. As has been mentioned, the rural population of the developing countries need high temperature heat mainly for cooking and therefore solar cookers with appropriate concentrators can meet this requirement. There are many shapes of solar cookers using various geometrical arrangements of concentrators. Among these, the box type cooker has been widely disseminated because of its compact design that makes it easy to transport. Others that use large parabolic concentrators are more efficient but more difficult to transport. There are also those very large cookers that are permanently installed for institutions such as schools and

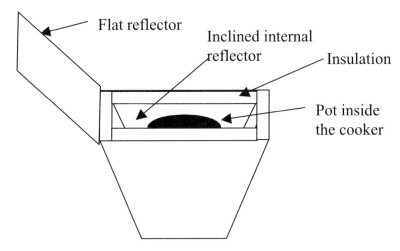

Fig. 5.4. Solar Box Cooker with a flat reflector

hospitals, which do bulk cooking for a large number of people. Such cookers usually use large compound parabolic concentrators.

Solar box cookers also come in various geometrical shapes. Figure 5.4 is just an illustration of one type but whatever the shape, they are not as efficient as those that use parabolic concentrators. The advantage is that they are convenient to use and transport. The top of the box is covered with a thin sheet of glass to prevent heat from escaping out of the cooking chamber. Figure 5.5 shows a more efficient cooker which is usually larger than the box cooker.

The concentrator structure can be mounted on a frame that allows it to rotate on a horizontal axis so that it can be adjusted to face the sun all the time. Inside surface of the parabolic dish in lined with a highly reflective material that redirects sunlight to the position of the pot. This type of cooker is capable of reaching very high temperatures that are suitable for cooking many types of staple foods in East Africa such as bananas, rice and mixtures of cereals that need boiling only. But it may not be suitable for preparing "Ugali" that needs to be stirred and thoroughly mixed even in the semi-solid form. The other disadvantage is that some of the reflected light may end up on the face and hands of the operator causing discomfort and suffering. Like box cookers, many of these parabolic cookers are already in use in many parts of the world as a result of promotional activities of a number of non-governmental organizations. Countries, which have made significant progress in promoting solar cookers are India, Pakistan and China. The African con-

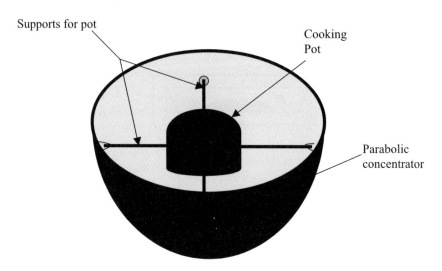

Fig. 5.5. Parabolic concentrator type solar cooker

tinent is still lagging behind but the potential for their application is definitely high and should be exploited. The cookers work well in direct sunlight and this means that they can only be used during daytime when there is direct sunshine. Another energy source is required for nighttime applications preferably for the final preparation of the food when pre-heating (initial cooking) has been done by the solar cooker. Except for the high quality reflector material, all the other cooker materials are cheap and locally available. The construction itself does not require specialized skills and local artisans can easily reproduce the cookers from well-designed samples. Unlike the box cookers, parabolic cookers can be scaled up to any desirable size. If it is necessary to use a large cooker, then it becomes necessary to design the cooker such that the focal point is far away from the concentrator. This will make it easy and safe to use since relatively high temperatures can be achieved from large concentrators. In this case, compound parabolic concentrators are used to redirect concentrated solar radiation on to the cooking pot suitably placed inside the kitchen. The temperatures reached and the generally large focal area of the reflector make it possible to use large cooking pots that can prepare food for many people. Such cookers, capable of making food for up to 500 people, have been designed and tested in East Africa. The Appropriate Technology Center at Kenyatta University in Kenya has the capability to design, construct, test and install these compound parabolic solar cookers, sometimes referred to as institutional solar cookers. A typical compound parabolic solar cooker is shown in Fig. 5.6. Government institutions such as schools, hospitals, village polytechnics, institutes of technologies and prisons would be the largest users of these large cookers and therefore for them to succeed, it is important for the governments to positively promote their use through effective encouragement of the institutions to invest in them.

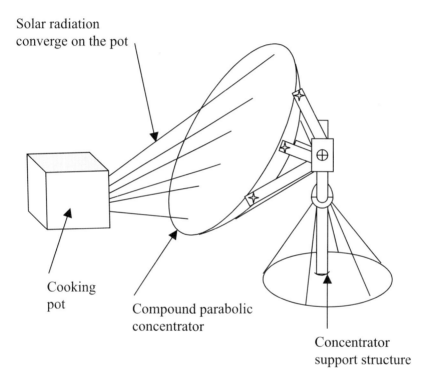

Fig. 5.6. Compound Parabolic Solar Cooker (Institutional Solar Cooker)

The cooking pot is suitably located in the kitchen where concentrated solar radiation passing through the kitchen window can reach it. Often, the efficiency is improved by placing additional reflector at the bottom of the cooking pot, which redirects the already concentrated solar radiation to the bottom of the pot. Experience has shown that heat from the bottom of the pot is transferred within and around the pot more effectively than when only one side of the pot is heated.

Cooking in most of the institutions is done during the day and so these cookers can be very useful. Obviously additional source of fuel would be necessary and indeed should be available all the time to be used on a cloudy day. Usually a lot of heat is required to boil water for various applications in the institutions and solar cooker can easily accomplish this. Since solar cookers can work only during the daytime, they can be used for boiling water and pre-cooking purposes in order to conserve biomass and kerosene fuels. These can then be used only briefly during the final preparation or warming of food at night. Foods like maize, peas, beans, bananas, potatoes and vegetables some of which take several hours to cook, require straight boiling with very little attention. Such foods are common meals in the East Africa and can be adequately cooked using solar cookers and this would lead to substantial savings on biomass and fossil fuels.

5.3 Solar Electricity

Sunlight can be directly converted into electricity using specially produced semiconductor in which the energy barrier between the conduction band and the valence bands is in the order of photon energy so that the light photon can excite electrons to jump across the energy barrier onto the conduction band where they become 'nearly' free to move and hence create current. This phenomenon is known as photovoltaic effect and the electricity produced is referred to as PV power. The process of producing semiconductors of high grade and processing it to finally become a solar cell that readily converts sunlight into electricity is very delicate and expensive. It requires expensive high precision equipment. For this reason solar photovoltaic cells have been too expensive for low-income communities. However, the advances, which have been made in PV cell production, are gradually making the technology become more cost-effective at commercial scale than was expected. It is hoped that within the next few years, solar PV power will be competitive and probably the most preferred energy source. These advances in photovoltaic technology including the development of thin film devices hold the key to new power supply systems, which could speed up rural electrification in developing countries. It is important to understand how PV cell evolved in order to appreciate their potential applications.

5.3.1 The Development of Photovoltaic

Photovoltaic effect was first observed in the second half of the 19th century but the selenium material which was used could only convert about 0.1 - 0.2 per cent of the energy in the incident sunlight into electricity. Attempts to improve this did not yield any hope for their use as a power source. Moreover, selenium, which is a copper-ore- based element, was too expensive. It was not until 1954 that a team of American scientists discovered that a silicon device, which they were testing for a completely different purpose, produced electricity when exposed to sunlight. They went ahead and made more efficient solar cells from silicon with a conversion efficiency of about 1%. This generated a lot of excitement because, unlike the more expensive selenium, silicon happened to be one of the second most abundant elements with a share of about 28% of the earth's crust. By the end of the 1950s, the conversion efficiency of silicon solar cells had been increased to just above 10%. At the same time the first application of photovoltaic as a significant source of electricity was introduced in pioneer space vehicles. Since that time, photovoltaic has been used in space satellites for communication and space exploration in general. For the satellites, the main factors in the application of solar cells were their reliability, low weight, durability and radiation resistance. High cost was not of immediate concern since space technology, in general, involved high expenditures. The first solar powered satellite, Vanguard 1, was launched in March 1958 with silicon solar cells to power its 5 mW back-up transmitter. A couple of months later, the then USSR launched a much larger solar-powered satellite. Since then solar power has been used practically in all spacecrafts. It would appear that if it were not the timely invention of solar cells, space exploration would not have progressed at the rate it did. In fact, photovoltaic was originally intended for use exclusively in spacecraft. Its terrestrial application was only later recognized when, in 1973, major oil producers in the Middle East threatened to stifle world economy by restricting oil supply. Sudden attention was then given to solar cell, and this encouraged the developing countries to also consider it as an alternative energy supply. They saw in solar cells the possibility of supplying power to millions of their population who had not been connected to the national power networks due to the high cost of expanding national grids to cover remote and isolated rural areas. Consequently, research on solar cell materials was globally intensified. Today, in addition to silicon, several new materials with higher solar energy conversion efficiencies have been developed. Some examples of these are Gallium Arsenide (GaAs), Gallium Indium Phosphate (GaInP2), thin film cells based on amorphous silicon (a-Si:H), crystalline silicon, Copper Gallium Indium Diselenide(CuGaInSe$_2$ or CIGS), Cadmium Telluride(CdTe) and Cadmium Sulphide (CdS). More recently, new materials have been investigated as possible candidates for solar cells. Some of these such as single crystal C$_6$0 and polycrystalline CdSe on tin substrate are used in photo electrochemical solar energy conversion to produce liquid solar cells (photo electrochemical solar cells). A few other materials are currently being analysed for their solar energy conversion characteristics. Among the chemical methods of converting solar energy, those involving semiconductors in the form of suspensions, monocrystals, polycrystalline or amorphous materials and using decomposition of water and production of hydrogen have yielded some promising results. Thus the material base for solar cell technology continues to expand. However, the costs associated with the initial investment in solar energy systems have remained some of the major prohibiting factors in their rural applications even though they have been confirmed as a commercially viable sources of electricity.

5.3.2 Photovoltaic Cost Reduction Trends

The cost of producing a solar cell is mainly determined by the cost of the material and the technique used in the manufacturing process. Other related costs include the inherently low efficiencies of, and the increased demand on, the right grade of the production material. There is therefore a need for further research and development, which would result into cost reduction to a competitive level with other well-established power generating facilities. Consider, as an example, solar cell made out of silicon which is the most abundant and perhaps the cheapest solar material: metallurgical grade silicon with impurity level of about 1% has to be processed to an impurity level of about 10^-6 per cent to raise it to a solar grade silicon, which costs US$ 15 - 20 per kg. To use this silicon to make a solar cell, it must be transformed (melted and then grown) into either a single crystal or multicrystalline ingot before it is cut into thin slices (wafers) of about 300mm, which eventually become the solar cells. This process wastes about 50% of the material. Furthermore, in order to obtain the required biases (p- and n-types), the wafers are doped with another element. At the end of the process the price of silicon has increased to about US$ 50 per kilogram while the efficiency remains quite low at about 13%. Some improvements have been made in the wafer slicing techniques to reduce material waste. This led to a reduction of the cell cost, but still not to the desired level. Fortunately the ingot-wafer process is not the only method of making solar cells. Another method using thin film technology is steadily gaining popularity. Instead of growing an ingot, silicon or any other suitable material can be deposited as a thin film onto a substrate and, after appropriate conditioning, a solar cell can be produced. This technique has several advantages over wafer method such as material economy, choices of sizes and shapes of the cells and a wide range of design flexibilities. There are also many techniques of tailoring materials in this form for photovoltaic applications. Some of the common methods are:

1. Plasma Chemical Vapour Deposition
   - High frequency glow discharge
   - Direct current glow discharge
1. Vacuum Evaporation Method
   - Electron beam evaporation
   - Reactive laser evaporation
2. Sputtering
   - High frequency sputtering
   - Magnetron sputtering
3. Pyrolysis
   - Normal pressure chemical vapour deposition
   - Low pressure CVD
   - Homogeneous CVD
4. Photo Chemical Vapour Deposition

These techniques require high precision machines and equipment, which have proved to be too expensive for the developing countries. They also use a variety of materials some of which may not be readily available. The materials range from single element semiconductors (e.g., silicon) to compound semiconductors (e.g. Gallium Arsenide). The advances in thin film techniques and solar materials research has been important in semiconductor technology, which contributed to the progress made in microelectronics used in various communication and data processing devices. Mass production of solar cells and the use of new efficient technologies in the manufacturing process have, over the years, reduced the cost of photovoltaics to affordable levels. But to make photovoltaic competitive with other sources of electric power, many manufacturers around the world are continuing research and development in order to achieve specific cost objectives which will provide electrical output in the US\$ 0.03 - 0.05 per kilowatt-hour range. This will make the cost of solar power competitive with conventional fossil fuel costs. Studies of PV systems with reference to their application in Africa indicate that the ratio of the cost of a PV system to the total generated energy in its normal lifetime ranges from US\$ 0.10 to 0.15 per kilowatt-hour. This ratio is expected to decrease in future and is an indication that photovoltaic home systems will become the most ideal energy for unplanned randomly scattered homes and villages, which is the typical situation of African rural settlements. It has also been reported that the worldwide average cost of grid electricity is about US\$ 900 per household. But in unfavourable cases, such as those prevailing in Africa, with long distances, difficult terrains and low population densities, power line for one household costs a lot more. In this case, grid extension costs about US\$ 10,000 per km. On the other hand a solar home unit, large enough to power a radio, black-and-white television and three light points, would today cost less than US\$ 500.

5.3.3 Properties of Solar Cell Materials

The conversion of sunlight directly into electricity is based on the band theory of solids, which considers a crystal as an assembly of atoms arranged in a regular pattern to form the solid substance. Atomic nuclei in this orderly arrangement can vibrate at a fixed point without moving to other positions in the crystal. Electrostatic force keeps electrons closely attracted to their nuclei but the outermost electrons can behave as if they do not belong to any particular atom, but to the crystal as a whole. A material with such 'nearly' free electrons is a good conductor of electricity because any additional energy supplied to the material in the form of electromotive force will cause electrons to move. However, some materials do not have these free electrons but it is possible to create them by supplying some amount of energy to the material. This means that electrons are in some defined energy states and to change their state so that they become 'free', addition energy from an external source is required. Some materials, known as semi-conductors, require just a small amount of additional energy while others that require large energies are known as insulators. Semiconductor property is more suitable for the conversion of sunlight photon energy into electricity through a process in which the photon interaction with electron inside a material excites the electron to jump from its energy state to a state for free electron. To make the material to operate as a solar cell, there must exist a condition, which prevents the electrons from internally returning to their original energy states. The problem is that there are no natural materials with these properties and therefore suitable semiconductors must be artificially made for solar energy conversion purposes. We shall consider a process of making such semiconductor from silicon. Silicon can be obtained from sand and the convenient material to start with is silicon dioxide in the form of white quartzite sand, which is melted in an electric arc furnace. Carbon arc reacts with oxygen in the silicon dioxide to form carbon dioxide and molten silicon. The obtained molten silicon is of metallurgical grade with 1% impurity but still not pure enough to make solar cells. It has to be purified further to semi-conductor grade silicon and this is achieved by thermal decomposition of silane or some other gaseous silicon compound. A seed rod of pure silicon is heated red-hot in a sealed chamber. Purified silicon compound is then admitted into the chamber and when the molecules of this compound strike the pure silicon rod, they break down to form elemental silicon, which builds up on the rod as a pure silicon crystal. When the rod has grown to the desired size, it is removed and is ready to be made into single-crystal solar cells. Another silicon purification method is known as zone refining in which a rod of metallurgical grade silicon is fixed in a machine with a moving induction heater coil, which melts a zone through the entire rod, starting at the bottom and slowly moving to the other end. As the coil moves, the front edge of the front edge of the melting zone continues to melts while the rear edge solidifies. The impurities are excluded from the solidifying zone and are dragged along by the melting zone so that, finally, the rod is pure except the last end, which can be discarded. Both these methods require controlled environment and accurately maintained uniform process rate. The actual process of making a solar cell now begins with the obtained pure silicon. First, the pure silicon ingot must be sliced into thin slices known as wafers, and then a junction must be created in the material to separate the electrons from the holes so that internal recombination process does not occur. On one side of the cell, we should have a p-type of material and on the other, an n-type so that one donates electrons while the other accepts them. The actual junction must be a region where there are no electron or holes and is sometimes referred to as the depleting layer. The creation of p- and n-types semiconductors is done through a process known as doping in which a chosen impurity e.g. boron or phosphorous is added into the silicon very carefully and in a well-controlled manner. When forming a junction, usually p-type silicon is obtained by adding boron atoms during crystal growth. Later, phosphorous atoms are added on top of the p-type silicon such that there are more phosphorous atoms than boron. This process will produce an n-type layer and a p-n junction is formed. This process can be achieved in a high temperature diffusion furnace having a gas of the dopant. Dopant atoms strike the surface of the semiconductor and then slowly diffuse into the material. If the temperature and the exposure time are properly controlled, a uniform junction would be formed at a known distance inside the cell. But this is not the only method of creating a p-n junction, there are other methods including bombardment of the silicon with ions of the dopant. In this case, the desired penetration depth is achieved by controlling ion speeds. Having made the cell, the next step is to connect the cells together and also provide leads to facilitate the completion of the electric circuit. All points joining one component to the other must be ohmic contacts that do not impede the flow of electrons into or out of the cell.

The p-type and n-type semiconductors form the structure represented in Fig. 5.7, such that the Fermi levels line up. When this is done, there emerges a difference in the vacuum zero levels for the two types of semiconductors. This represents the built-in potential that now exists because of the difference in the work functions of the two semiconductors. When sunlight falls on the cell as indicated, those light photons with energies greater than the band gap excite electrons from the valence to the conduction band and create hole-electron pairs. The electron, having been excited into the conduction band, moves into the depletion layer and 'slides down hill' towards the front of the layer. The result of this movement is equivalent to a hole 'rising up-hill' like a bubble into the p-layer. The accumulation of electrons and holes on the opposite sides of the depletion zone is made possible by the electric field that exists in the zone and this is the process that makes it possible for current to flow through the cell in one direction. Electron-hole pair created away from the junction will, due to the junction field, drift towards the junction and cause current flow as described above. The junction is simply a barrier between electrons and holes.

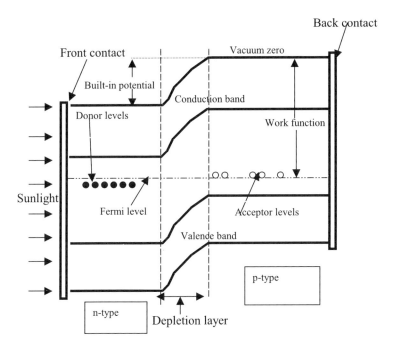

Fig. 5.7. A simplified representation of a solar cell

A barrier made by doping the same material like the one described above is known as a homojunction. There are other types of barriers that are normally used for solar cells. Some of the common ones are heterojunction, heteroface junction and Schottky barrier. A homojunction is a p-n junction of the same semiconductor material. Heterojunction is also a p-n barrier created between two different semiconductor materials one of which may have a larger energy gap than the other. Heteroface junction is similar to a homojunction but with an additional window layer of a larger band gap while the Schottky barrier is a metal/semiconductor junction.

5.3.4 The Solar Cell

We have discussed some of the processes of developing material properties that are required in order to convert sunlight directly into electricity. The most crucial aspects are;

(i) the creation of a barrier layer or junction and,
(ii)the energy gap between the valence and conduction bands.

These are the features that determine the conversion efficiency of a cell as we shall soon find out. First we need to know how much of the incoming solar radiation participates in exciting electrons from the valence to the conduction band so that they become 'free' to move and hence create current. The energy in the rays of sunlight is limited to the photons with wavelengths ranging from 0.2 to $3\mu m$, which correspond to an energy range of about 10 to 0.6 eV. The intensity however is very low at both ends of the spectrum particularly in the infrared region (Fig. 5.8). The energy of the radiated solar photon is given by

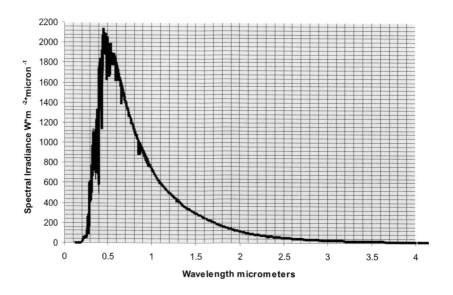

Fig. 5.8. Typical clear atmosphere solar radiation spectrum

$E = h\nu = hc/\lambda$, where h is Planck's constant, c is the speed of light and λ is the wavelength. The relationship is used to determine the photon energy that is likely to excite electrons to cross from the valence to conduction bands. From Fig. 5.8, it is clear that most of the radiation fall within wavelengths from 0.4 to 0.8μm with most photons having energies ranging from about 2 to 3 eV. This means that the cell energy gap (energy gap between conduction band and the valence band) must be less than 3 eV if sufficient quantities of electrons are to be excited by the light photon to produce some significant current. It is immediately clear that some of the photons in sunlight will not be able excite electrons because the energy gap of the cell is larger than the photon energy. Small band gaps increase reverse current and decrease the voltage from the cell and therefore they are not suitable. On the other hand, large energy gaps decrease reverse current and increase the voltage obtainable from the cell, but photo-generated current is reduced. Experience has shown that band gaps of 1.4 to 1.5 eV are the most suitable for solar radiation spectrum. This, however, is an ideal gap range, which is not easy to achieve. Therefore light photons with wavelengths greater than 1.0μm, and hence energies below this range, cannot take part in the conversion process. In reality, a lot more light photons than this do not participate in the conversion of sunlight into electricity. Some light rays are reflected while some do not strike electrons in the cell. There is also recombination of some excited electrons with the holes. All these processes significantly reduce the conversion efficiencies of solar cells and are the reasons why it is difficult to produce PV cell with efficiencies as high as 40%.

5.3.5 Recombination of Electrons and Holes

We shall now consider recombination process in more details. When an electron is excited and moves to a higher energy level, it leaves a hole behind and would be unstable in its new state. The tendency for this electron is to drift and eventually go back to its more stable state. The time an electron takes before it recombines is known as relaxation time,τ, having moved a typical distance (diffusion length),L, through the crystal lattice. If the material is pure, the relaxation time can be as long as one second. For a doped material, this time is short and may range from 10^{-2} to 10^{-8} seconds. The probability per unit time of a carrier recombining is

$$p(\tau) = \frac{1}{\tau} \tag{5.7}$$

For n electrons, the number of combinations per unit time is n/τ_n and for p carriers it is p/τ_n. At equilibrium in the same material, these re-combinations should be equal

$$\frac{n}{\tau_n} = \frac{p}{\tau_p} \tag{5.8}$$

Which means that

$$\tau_n = \frac{n}{p}\tau_p \tag{5.9}$$

and

$$\tau_p = \frac{p}{n}\tau_n. \tag{5.10}$$

Thermally generated carriers diffuse through the lattice down a concentration gradient dN/dx to produce a current density, in the direction of x,

$$J_x = -D\frac{dN}{dx}, \tag{5.11}$$

where D is the diffusion constant.

Within the diffusion time, τ, the diffusion distance, L, is given by Einstein's relationship:

$$L = (D\tau)^{1/2}. \tag{5.12}$$

For a typical diffusion length for minority carriers in a p-type silicon for which $D \approx 10^{-3}$ m^2s^{-1} and $\tau \approx 10^{-5}$s, the diffusion length is $L = \sqrt{(10^{-3}x10^{-5})} = 10^{-4}m = 100\mu m$. It should be noted that $L >> w$, where w is the junction width of a typical p-n junction making it possible for most carriers to drift to the junction and even go through it before recombining.

5.3.6 Cell Efficiency

The power of the cell is determined by the product of its current and voltage, which mutually depend on each other as shown in Fig. 5.9. Since power is given by the area under the curve, a powerful cell must have as large an area as possible. Both current and voltage should therefore have high values whose product gives the highest possible area. This is the peak power of the cell. Evidently, the area given by the peak power does not fill the total area enclosed by the current-voltage characteristic curve because peak power occurs at a combination of voltage and current that are lower than open circuit voltage and short circuit current respectively.

The highest possible current is the short circuit current, I_{sc}, while the maximum voltage is the open circuit voltage. Maximum power is

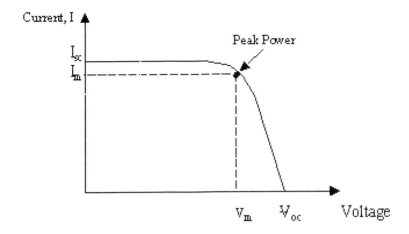

Fig. 5.9. The Current-Voltage characteristics of a solar cell

when the product of current and voltage is maximum and is represented by a point on the curve marked I_m and V_m. The efficiency of the cell is therefore defined as this product divided by solar radiation incident on the cell. A factor known as the fill factor, ff, is given by the ratio of the peak power to the product of the short circuit current, I_{sc}, and the open circuit voltage, V_{oc}.

$$ff = \frac{I_m V_m}{I_{sc} V_{oc}}. \tag{5.13}$$

The conclusion we can make from these solar conversion parameters is that high efficiencies of the cells would be difficult to achieve without radiation concentration mechanism. At present operational or module efficiencies are in the range between 7 to 14%. However, laboratory cell efficiencies of double this range have been achieved. It is of course necessary to understand these theoretical analyzes of solar cells to be able to follow the discussion on practical production methods one of which is given in the next section.

At the application level, solar photovoltaic panels provide the user with a number of options but the most important aspect is that the user can choose to gradually build the system from a small unit to a large while using it. This means that one can start with a small system that supplies power to the lighting system only. After saving more money, additional panel and battery can be purchased so that there is enough power for both lighting and a radio or a television. More components can be added as money becomes available until finally the user can power all household appliances including lighting. This modular nature of photovoltaics systems is an aspect that can effectively take care of the generally low income of the rural communities if they are encouraged to "begin small" and grow at one's own pace without sacrificing other family requirements.

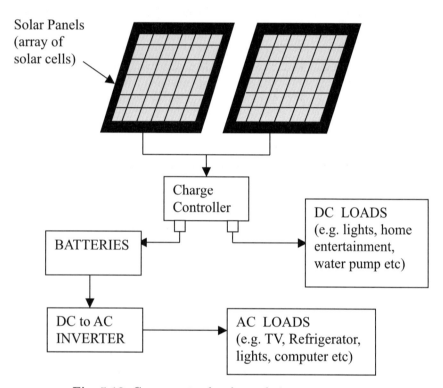

Fig. 5.10. Components of a photovoltaic power system

A working solar electricity generator must have the panel, which is an array of solar cells and is the source of electricity. When sunlight reaches the cells, electrons are forced to move from their low energy levels to higher levels inside the cells creating current through a suitably designed circuit. The flow of current will only continue as long as the panel is exposed to sunlight and this means that an electricity storage system must be provided for night use. The batteries fulfill this requirement; they are charged during the day and at night they are used to provide the required electricity. A charge controller is connected between the batteries and the panels to prevent the batteries from overcharging and hence protect them and prolong their lives. Both the batteries and the panels produce direct current but most modern household appliances operate on alternating current. For this reason a DC to AC inverter may be used to convert some of the battery's DC voltage to AC voltage in order to power those appliances that use AC. If the system is used for lighting only, then there is no need to have an inverter and if the battery storage is large compared to what the panels can produce at any given time, then the use of a charge controller may not be necessary. Thus the basic mandatory components are the panel and the battery. The high cost of the panel has been the major prohibiting factor in the use of solar electricity systems. However, as the costs of solar panels continue to decrease and, given this up-grading feature, solar electricity will be the best option for rural electrification. Furthermore its low voltage makes it safe to use in most rural houses in their present conditions. More than 90% of rural residential buildings are not suitable for high voltage electricity sources and this is a fact that has not been addressed by national grid power providers. Water leakages and structural damages, which are common occurrences in many residential buildings would have grave consequences if connected to high voltage national grid power. In fact the rate of power interruptions and power related deaths would increase sharply. Solar electricity is the solution to all these. The advantage that solar systems already have in East Africa is that the market is already well developed and practically all system components are available in the major towns. They are already in use in a number of remote establishments such as schools, missionary centers, hotels, dispensaries and other health facilities where they are used to power vaccine refrigerators, telecommunication facilities, water pumps, lights etc. The number of households using solar PV systems is also steadily growing. Besides photovoltaic technology, there is another method of generating electricity using solar energy heating characteristics. This involves the use of large mirrors to concentrate sunlight onto a receiver that heats gas and generates steam for turning a generator turbine. While this should also be considered, more attention should be given to production of electricity from photovoltaic solar cells.

5.3.7 Solar Cell Production Method

In order to appreciate the complexity of making solar cells and properly understand how solar cells work it is important to know about the

Fig. 5.11. Solar PV panel in use in Lake Victoria region of East Africa

nature of sunlight, its radiation spectrum and possible changes due to
a number of atmospheric factors. The next equally important knowledge is the understanding of how materials react to sunlight. This is
covered by modern solid-state theory and the band model of semiconductors, which give information on how the photon energy of sunlight
can be transferred to electrons in the solid material so that they acquire
enough energy to move from one area to another inside the material.
The electrons occupy various energy levels in the material and therefore only some of them are able to get enough energy from sunlight
to move. Thus, in terms of solid-state theory, there are many factors
that influence solar cell efficiency. Some of these include the size of
band gap, electron-hole recombination centers and reflectivity of the
material. However, the knowledge of materials science and the nature
of radiation from the sun are just the basic requirements. The process
of making solar cells from suitable materials is more complicated, energy consuming and expensive. This section gives, as an example, one
technique for making solar cells from silicon. Silicon is one of the most
abundant, and relatively cheap materials for making solar cells. It is
present in most rocks and minerals but the convenient starting material
is silicon dioxide in the form of white quartzite sand. This sand can be
reduced to silicon by melting it during which oxygen in silicon dioxide
reacts with the carbon to form carbon dioxide and molten silicon. An
electric arc furnace can be used to achieve this reaction, which produces

metallurgical-grade silicon with about 10% impurities. Such silicon has many applications in steel and other industries. However, its level of impurities is still too high for use in electronics or solar cell industry. The next stage is to produce semiconductor-grade silicon by adding a small percentage of dopant atoms in the silicon crystal lattice. In order to achieve this, the number of impurity atoms in the molten silicon must be small compared to the number of dopant atoms. This means that the silicon must be hyper pure and the most common method of producing this quality of silicon is by thermal decomposition of silane (or gaseous silicon compound). A seed rod of ultra pure silicon is heated red-hot in a sealed chamber in which a purified silicon compound is admitted. When the molecules hit the hot pure silicon, it is broken down to form elemental silicon which builds up on the pure silicon rod up to the desired size and then removed.

Another purification method known as zone refining uses a rod of metallurgical-grade silicon with a moving induction heater coil. The coil melts the rod progressively as it moves from the bottom of the rod to the top. Silicon that melts at the top edge of the zone solidifies at the bottom edge. As the liquid solidifies, impurities are excluded from the new crystal lattice and the molten zone picks and sweeps them to the top of the rod from where they can be discarded. Both these purification methods produce semiconductor-grade silicon but are expensive making and so the resultant product is also expensive. This is, however, the material that can be used to make solar cells but it has to be processed further in order to acquire the required properties for an electric cell.

The boundaries between silicon crystals act as traps for electrical current and therefore these boundaries should be removed. This means that the malticrystalline silicon should be grown into a single crystal without any boundaries. This can be done by dipping a seed crystal into a crucible of molten silicon and slowly withdrawing, pulling a large round single crystal as the molten silicon solidifies. This process is carefully controlled under predetermined withdrawal speed and temperature of the molten silicon to produce a crystal of desired size which would also depend on the size of the pulling machine and the amount of silicon the crucible can hold. This method of growing crystal is known as the czochralski process.

A small amount of dopant material can be added to the single silicon crystal during its growth in order to produce the required electronic properties. Usually boron is used as the dopant so that a deficiency of electrons is created in the crystal to make it a p-type semiconductor.

Since the cell is "energized" by sunlight, which can only penetrate a small depth of the silicon, the cell must be extremely thin. The bulky single crystalline p-type silicon material must be sliced into very thin wafers of thickness in the order of 300-500 microns. A special cutting machine with cooling lubricant is used to produce the slices (wafers). This method is obviously wasteful and a significant mount of silicon is converted into dust. The surface of the wafer is then polished using fine abrasives or chemical etching. A number of various surface treatments have also been developed. Finally the top half of the silicon semiconductor is made n-type by incorporating more phosphorus atoms than boron atoms. Excess phosphorus atoms at the top layer and boron atoms at the Bottom layer creates a p-n junction in the material with excess electron and holes in the phosphorus and boron layers respectively. The addition of phosphorus is done by placing wafers in a diffusion furnace and heating them to a high temperature in the presence of phosphorus gas. The process is carefully controlled under suitable temperature and time of the exposure in order to form a uniform junction at a known depth inside the wafer. Normally wafers are sealed back to back so that the backs are protected from receiving diffusing phosphorus atoms. There are other methods of forming front junctions e.g. by ion implantation in which the penetration depth is determined by controlling the speed of the ions hitting the wafer.

The formation of the junction completes the solar cell manufacturing process and what remains is to provide the circuit through which electric current can flow. Since single cells are very small and have relatively low voltage, it is necessary to join several cells together to produce the required peak power. The cells can therefore be connected in series or in parallel or a combination of these in order to produce the so called PV module of the desired peak voltage. Since the cell's potential difference exists between the front and back sides, the connections to the circuit have to be done at these surfaces. Both the front and back contacts must adhere very well to the cell and must be ohmic contacts with as low resistance as possible. The cell should receive as much direct sunlight as possible and, in this regard, the front contract should not prevent sunlight from reaching the cell. On the other hand, the back contract can be a solid metal coating, since light does not have to pass through this surface. In fact a reflective coating to reflect light back into the cell would be better. There are many methods and materials used in producing back and front ohmic contracts.

Silicon out of which the cell is made can reflect as much as 35% of the sunlight falling on it, reducing the amount of light that could have generated electricity. This property makes it necessary to coat all cells with anti reflection coating, which must be highly transparent to allow sunlight to reach the cell and initiate flow of electric current and therefore they are usually very thin, typically less than 0.1 micron (100 nm) thick.

There are many materials used for antireflective coating including silicon monoxide, silicon nitride and titanium dioxide and can be applied to the silicon cell using appropriate vacuum coating process such as vacuum evaporation or sputtering. Some high quality antireflection coatings are made up of three or more layers (multiplayer) and can reduce reflection down from about 35% to less than 2%. Thus, in general, the silicon solar cell is a pair of p-and n types semiconductors with the following

Characteristics: Junction that separates electrons and holes; ohmic front and back contacts and antireflection coating. Using the front and back contacts, several cells can be connected together to form a photovoltaic (PV) module or panel of desired peak power.

Other methods of making solar cells including cost reduction attempts and operating characteristics are discussed in Sect. 5.3.2. Fig. 5.11 shows a PV system used by Lake Victoria community in East Africa.

Solar electricity obtained from PV panel can be used for a number of purposes. Its application is especially appreciated in circumstance for which electricity from a centralized facility is not convenient or safe to use. Some of the examples of application of PV electricity include water pumping, street and domestic lighting, provision of light for fishing of Lake Victoria sardine at night and powering domestic appliances. One outstanding features of solar electricity is that it can be generated where it is needed, enabling many households far from grid power to use electric appliances including vital communication devises. Solar cell modules are reliable, quiet and most economical when supplying small amounts of power for lighting, radios, televisions, and a variety of small machines. Using solar cell modules as source of power for electric fences is rapidly becoming popular in the wake of increased insecurity and the need to effectively confine wild animals to designated game parks.

5.4 Wind Energy

Like solar, no country can claim to have no wind. It is everywhere but the only problem is that it is present in varying intensities that also depend on the seasons of the year for the particular location. Harnessing wind energy thus requires a n accurate assessment of the annual wind regime for the location. East Africa has suitable wind regimes for various applications such as water pumping, grain grinding and electricity generation. wind machines that are installed in the region are working well - a clear prove that the region's wing energy can be successfully harnessed. The region also has reputable wind machine manufacturers with guaranteed after-sales service and maintenance. However, these manufacturers are too few and cannot effectively cover the whole region. Furthermore, the near-monopoly situation also makes their products too expensive and, wind turbines being heavy and bulky, are too costly to import. There is need to develop local capacity in the design, manufacture, installation and use of wind machines.

5.4.1 Wind Energy Technology

Wind is air in motion and therefore has kinetic energy, which can be tapped and converted into other forms of energy. To do this, appropriately designed machine is needed to convert linear motion of the wind into a rotary motion, which can be linked to a generator to produce electricity. Wind energy has been harnessed for over 5,000 years when the Egyptians used it to sail their ships on river Nile. The Persians (area now occupied by Iran) later used wind energy to grind their grain. In the American continent, the earlier European settlers used windmills to grind grain, pump water and also cut wood in timber factories but it was not until the 17th century when Holland used wind energy in large scale to power industrial processes. As the technology was perfected, wind became an important source of electricity for rural communities. But as knowledge of generating electricity from other sources increased and more effective transmission technologies developed, the use of wind also declined and more technologies were developed on the basis of the new sources such as oil fuels and nuclear energy. However, the world soon realized that over reliance on oil was risky as it could be depleted in future. In addition, its use was strongly linked to undesirable environmental impacts, which needed to be checked and so attention was again turned to the search for alternative sources of energy and wind became one of the important alternatives. Today, wind generators are a common site in Germany, the leading wind energy nation and also in the USA, Denmark, India, Spain, Holland and UK. Countries like China, Sweden and Canada also use a significant number of wind machines to generate electricity. Wind energy is converted to electric energy in two main stages. The first stage is to use the kinetic energy of the wind to turn a turbine. The circular motion of the turbine is then used to rotate the coils of conducting loops of wire inside a magnetic field in order to produce electricity. The rotational speed of the wind turbine is not constant since it is directly proportional to the wind speed which varies from time to time. At some times, there may be no wind at all while at other times wind speed may be so slow that it cannot turn the turbine blades. The generator on the other hand requires high rotational speeds and so it is usually necessary to use gears between the turbine and the generator in order to achieve the required rotational speeds. The number of revolutions per minute (rpm) of a wind turbine rotor can range from 40 to 400 rpm while generators normally require rotational speeds ranging from 1,000 to 2,000 rpm and this is achieved through the gear-box transmission. The generator can produce either Alternating Current (AC) or Direct Current (DC) depending on whether rings or commutator is used to pick the generated voltage. Whatever the case, the generated electricity must be regulated to the right voltage and, in the case of AC, to right frequency to suit the final application. Some DC wind generators do not use gear transmission systems but this means that larger generators are needed for the same power output. These are the direct drive systems. In general, the blades of the wind turbine must efficiently use the power in the wind to turn and therefore each blade must not disturb the wind for the next blade but should be long enough to capture as much wind as possible. The minimum interference requirement puts a limit to the number of blades that a turbine can have and so most wind turbines have only two or three large blades. Some large wind machines have blades with sweep diameters of 20 meters or more mounted on towers rising to 70 meters and above. However, the major design considerations are related to the required speeds for the prevailing wind regime in the area and the type of wind machine that would suit the final application. There are only two main classes of wind machines: vertical axis and horizontal axis. A vertical axis machine rotates on an axis that is perpendicular to the ground while horizontal machine rotates on an axis that is parallel to the ground. There are a number of available designs for both types. However, due to some operational requirements, vertical axis machines are not commonly used but they are available. The basic vertical axis designs are the Darrieus, the Savonius, the Musgrove, the Giromill and Evans types. The widely used types are the horizontal axis types in which the blades can be upstream or downstream of the wind flow in relation to the position of the tower. If the wind reaches the blades before passing the tower, then it is an upstream machine. In this case it is usually necessary to have a tail vane to turn the blades to face the wind all the time. There are many different mechanisms for turning the blades to position themselves for maximum power extraction. Crucial design considerations for these machines are the cut-in speed and the cut-out speed. If the wind speed is too low for the machine to generate any usable power, then it should stop rotating or rotate but not generate power. As soon as the wind speed picks up above a certain minimum, the blades should turn and generate power. This minimum speed is the cut-in speed for the turbine and is typically between 3 and 5 ms-1 depending on the size of the turbine and power generation range. On the other hand, wind speed can be so high that it can damage the machine and in this case the turbine shuts down and cease to generate power. The speed at which this occurs is called the cut-out speed and is typically between $20ms^{-1}$ and $40ms^{-1}$. This can be achieved in a number of ways such activating automatic brake using wind speed sensor, or pitching the blades to pick less wind, or by activating special mechanisms to turn the blades away from the wind. Normal wind machine operation then resumes when the wind speed drops down to a safe level. The other design factors are presented in the following theoretical analysis. The theory of wind energy conversion is best understood from the basics of Linear Momentum Theory.

In the unperturbed state, wind has kinetic energy per unit time (power), P_o, which can be expressed in terms of its mass flow rate.

$$P_o = \frac{1}{2}\rho A_1 u_o u_o{}^2 = \frac{1}{2}\rho A_1 u_o{}^3. \tag{5.14}$$

This is the power in the wind for an ideal case. A_1 is the cross section area and u_o is the undisturbed wind speed. The real case is different since the presence of the turbine in the wind affects the immediate up-stream and down-stream velocities and therefore there is change of pressure. It is the magnitude of theses changes that determine how much power would be extracted from the wind. Consider a case in which the turbine is located at position one and covers a disc area, A_1 (Fig. 5.11). The thrust or force on the turbine per unit time is

$$F = \dot{m}u_o - \dot{m}u_2. \tag{5.15}$$

It is assumed here that the wind speed, u1, is uniform so that the power extracted by the turbine is

$$P_r = Fu_1 = \dot{m}(u_o - u_2)u_1. \tag{5.16}$$

This should be equal to the energy lost by the wind per unit time, which is given as

$$P_w = \frac{1}{2}\dot{m}(u_o^2 - u_2^2) = P_r = \frac{1}{2}\dot{m}(u_o^2 - u_2^2). \tag{5.17}$$

The velocity at the turbine, u_1, has been eliminated by the fact that when we equate power extracted by the turbine to energy lost by the wind, we find that this velocity is:

$$u_1 = \frac{1}{2}(u_o + u_2). \tag{5.18}$$

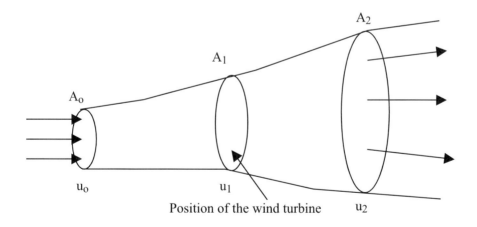

Fig. 5.12. Wind perturbation due to the presence of the turbine

According to this linear momentum theory, the air speed through the disc covered by the rotating turbine cannot be less than half the unperturbed wind speed since this speed, u_1, is more or less the average of the sum of the unperturbed upstream and down stream velocities. It can however, be equal to half the unperturbed upstream velocity if air leaves the turbine with zero velocity, a situation that is practically not possible because the wind must continue to move away from the turbine to prevent excessive back pressure that could stall the turbine.

Mass of air flowing through the area of the turbine per unit time is

$$\dot{m} = \rho A_1 u_1, \tag{5.19}$$

giving power extracted P_T as

$$P_T = \rho A_1 u_1^2 (u_o - u_2). \tag{5.20}$$

But u_2 can be expressed in terms of u_o and u_1 as: $u_2 = 2u_1 - u_o$, and if this is put in (5.20), then power extracted is

$$P_T = 2\rho A_1 u_1^2 (u_o - u_1). \tag{5.21}$$

We have seen that the presence of the turbine interferes with the speed of the wind both upstream and downstream. This interference must be included in the design considerations and therefore must be properly understood. It should also be remembered that the fact that wind speed through the turbine cannot be less than half of that of the unperturbed wind means that there is a limit to the power that can be extracted from the wind. Moreover enough power must be left in the wind to enable it to continue moving downstream. To analyze all these, let an interference factor, a, be defined as a fractional wind speed decrease at the turbine and use it to define the power extraction limitations

$$a = \frac{u_o - u_1}{u_o}, \tag{5.22}$$

which gives $u_1 = u_o(1 - a)$ so that $\frac{1}{2}(u_o + u_2) = u_o(1 - a)$. The factor, a, can now be expressed in terms of u_o and u_2 which can be easily measured.

$$a = \frac{1}{2}\left(\frac{u_o - u_1}{u_o}\right). \tag{5.23}$$

This factor, which is also referred to as the perturbation factor, is introduced into (5.21) to obtain

$$P_T = \frac{1}{2}\rho A_1 u_o^3 \left[4a(1-a)^2\right]. \tag{5.24}$$

The term outside the square parenthesis is obviously the power in the unperturbed wind at the turbine, that is, the maximum power that could be delivered to the turbine by the wind if the turbine did not interfere with the wind motion. Clearly the term inside the square parenthesis is the power extraction coefficient. Equation (5.24) is now written in terms of the power coefficient, C_p, and the unperturbed wind power, P_o as

$$P_T = 4a(1-a)^2 P_o = C_p P_o. \tag{5.25}$$

It is important to know the value of the perturbation factor, a, for which maximum power extraction would be achieved. This would occur at a suitable value of a when the gradient dC_p/da is zero - a condition that gives $a = 1/3$. This value of the perturbation factor gives the maximum value of the power coefficient; $C_p \approx 0.59$, and this is the Betz criterion. It means that only about half of the power in the wind can be extracted because wind must have kinetic energy to leave the turbine. The Betz condition applies to an ideal wind machine. Practical machines, however, convert less energy than this because of aerodynamic imperfections, mechanical and electrical losses. The actual power extracted would be less than that calculated using linear momentum theory and depends on the type and details of the design and operating conditions. One of the most important design considerations is the rated wind speed. That is, the lowest wind speed at which full power output is produced by the machine. The output at any higher speed than this is controlled to this level. Thus wind turbines having about 2 m length of turbine blade (sweep diameter of 4 m) would be rated between 1.5 to 5 kW depending on the actual rated speeds which are usually less than $14 m s^{-1}$. Medium size units with sweep diameters of about 15 m are usually rated at about 100 kW, while large turbines with sweep diameters of about 60 m are usually rated between 2 to 3 MW. Practically it is uneconomical to design for rated wind speeds much greater than 15 ms^{-1}. Since wind power is proportional to the cube of its speed, it is this speed that is of prime consideration in turbine design. Its annual average, distribution or frequency of occurrence is of great importance in assessing the energy potentiality of the site. Thus wind speed measurements should be analysed and presented in the form of a velocity-duration curve or velocity-frequency curve. The general forms of these curves are shown in Figs. 5.13 and 5.14.

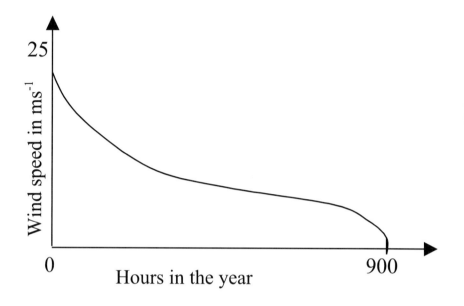

Fig. 5.13. Typical velocity-duration curve for a site

This curve shift the position up or down along the wind speed axis depending on the wind speed distribution at the site.

Velocity frequency curves have different shapes and maxima that depend on wind speed distribution. In the case presented in Fig. 5.14, the most frequent speed is 8 ms^{-1}. It is generally lower than the mean velocity and as the mean velocity increases, duration of most frequent velocity decreases but the area under the curve remains constant. Curves showing how the most frequent speeds vary with variation of annual mean speeds are also useful and represent different sites in a given region. All these information on wind regime for a particular area is important for the design of a suitable wind turbine. Rated wind speed is always between cut-in wind speed and furling wind speed. Thus variations of wind speeds for daily, monthly and yearly periods must be considered. Apart from wind regime considerations, there are other factors that influence the choice of a suitable site for wind machine. Such factors include accessibility to the site with heavy loads such as distance from either railway line or main roads, distance from electric-

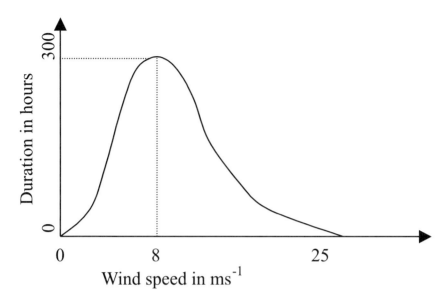

Fig. 5.14. Typical velocity-frequency curve

ity supply line or demand concentration. The nature of the ground also affects the cost of construction of the foundation.

As has been mentioned above, there are a number of factors that must be considered when designing a wind machine. More of these design considerations are discussed below including the factors that determine the number of blades a particular wind turbine should have. This is important because the presence of the turbine also interferes with the smooth flow of the wind.

The wind speed drop across the turbine causes pressure difference between the immediate upstream and the near downstream. Let us consider this pressure drop, $\triangle P$, and the axial thrust, F_T, (that is, the force along the axis of the turbine). The pressure drop is:

$$\triangle P = \frac{1}{2}\rho \left(u_o^2 - u_2^2 \right), \tag{5.26}$$

which gives maximum value when down stream velocity is zero. The maximum force on the turbine (the thrust) is

$$F_{Tmax} = \frac{1}{2}\rho A_1 u_o^2, \tag{5.27}$$

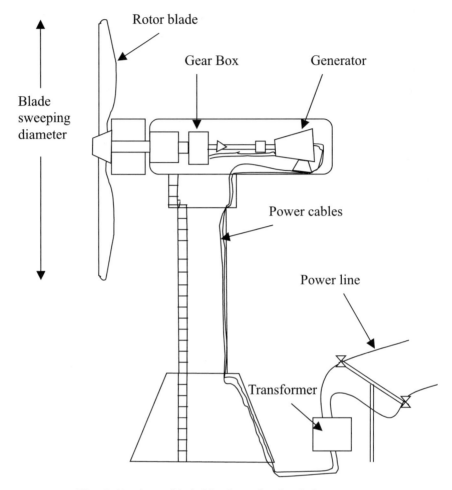

Fig. 5.15. A two-bladed horizontal axis wind generator

which is equal to loss of momentum of the wind when $u_2 = 0$. Under conditions when $u_2 \neq 0$, the axial thrust is

$$F_T = \rho A_1 u_1 (u_o - u_2). \tag{5.28}$$

Both u_1 and u_2 can be replaced by $u_1 = u_o(1-a)$ and $u_2 = u_o(1-2a)$ to obtain

$$F_T = \frac{1}{2} \rho A_1 u_o^2 \left[4a(1 - a) \right] = C_F F_{Tmax}. \tag{5.29}$$

The quantity outside the square parenthesis is the force hitting the area cut by the rotating turbine, and the quantity inside the square brackets is the coaxial force coefficient, C_F. The maximum value of coaxial force coefficient, C_F, occurs at a particular suitable value of a, when the gradient, $dC_F/da = 0$. It is found that this happens when $a = 1/2$ and C_{Fmax} assumes the value of one. This condition implies that $u_2 = 0$, a situation which is not permitted for the reasons given above. However, we already know that maximum power extraction is when $a = 1/3$, and this corresponds to $C_F = 8/9$. Thus for practical maximum power extraction, the coaxial force coefficient is not unity but 8/9. Note the difference between co-axial force coefficient, C_F and power coefficient C_p and that although their values depend on perturbation factor, a, they take different maximum values. All these factors and limitations must be considered when designing a wind energy conversion machine. There are other design considerations, which are not discussed at this stage. However, for small wind machines designed to generate electricity, the hub of the rotating blades is directly coupled to the rotor of the generator.

It must be remembered that the linear kinetic energy of the wind has to be converted into rotary motion of the turbine. This means that the wind must turn the turbine. Thus the torque causing rotational shaft power should be determined in order to ascertain whether or not a given wind speed would turn the turbine. Maximum torque would occur if maximum thrust is applied at the tip of the turbine blades furthest from the rotational axis. If the radius of the propeller turbine is R, then maximum torque, Ω_{max}, is:

$$\Omega_{max} = F_{Tmax}R, \tag{5.30}$$

giving

$$\Omega_{max} = \frac{1}{2}\rho A_1 U_o^2 R. \tag{5.31}$$

We can now define a torque coefficient, C_Ω, for any working machine as

$$C_\Omega = \frac{\Omega_{max}}{\Omega}, \tag{5.32}$$

where Ω is the shaft torque of the machine (turbine). If the size of the turbine and the speed of the wind are known then, assuming maximum thrust and hence maximum torque, the angular frequency of the turbine can be determined. To do this, the concept of tip speed ratio, γ, is introduced. This is defined as the ratio of blade tip speed u_t to the undisturbed wind speed u_o

$$\gamma = \frac{u_t}{u_o} = \frac{R\omega}{u_o} \qquad (5.33)$$

Where ω is the turbine rotational frequency. Equation (5.31) can now be written as:

$$\Omega_{max} = \frac{1}{2}\rho A_1 u_o^2 \frac{u_o\gamma}{\omega}. \qquad (5.34)$$

Using (5.28), maximum torque can be expressed in terms of the power in the wind P_o so that

$$\Omega_{max} = \frac{\gamma P_o}{\gamma}. \qquad (5.35)$$

Shaft power which is derived from power extracted from the wind P_T, is related to rotational frequency and turbine shaft torque:

$$P_T = \gamma\Omega. \qquad (5.36)$$

It is clear that shaft power is equal to the power extracted from the wind so that

$$C_P = \gamma C_\Omega. \qquad (5.37)$$

Since maximum value of C_p is known, the maximum value of C_Ω can also be determined if γ is known.

Finally the design of a wind machine must consider dynamic matching. This is an attempt to 'fit' the machine into the wind regime. If t_b is the time a following blade takes to move into the position of the preceding blade and t_w is the time wind takes to re-establish its steady velocity after the machine interference, then dynamic matching is an attempt to make t_b equal to t_w. In effect, this means matching wind speed to turbine rotational frequency. It is now possible to relate tb to t_w to some distances:

$$t_b = \frac{2\pi}{n\omega}, \qquad (5.38)$$

where n is the number of blades and

$$t_w \cong \frac{d}{u_o}, \qquad (5.39)$$

where d is the length of the disturbed section of the wind. Since maximum extraction occurs when $t_b \approx t_w$, we can write:

$$\frac{n\omega}{u_o} \approx \frac{2\pi}{d}, \tag{5.40}$$

and since $\gamma = \frac{R\omega}{u_o}$, we can write

$$\gamma = \frac{2\pi}{n}\left(\frac{R}{d}\right)\frac{1}{\kappa}. \tag{5.41}$$

The ratio R/d should be equal to one for maximum power extraction so that for $R \approx d$ and $\gamma_o = \frac{2\pi}{n\kappa}$

Generally, practical machines show that $\kappa \approx \frac{1}{2}$ giving:

$$\gamma_o = \frac{4\pi}{n} \tag{5.42}$$

Using this relationship, we can determine the number of blades suitable for desired turbine speeds in a known wind regime.

The design of any wind turbine starts with the type of application for which it is to be used. This is important because of the different torques required for specific applications. For example, if the machine is to be used for water pumping, it must be able to produce the high torque required to turn the heavy mechanical couplings and the pump. On the other hand, if it is to be used for electricity generation, the machine should be able to rotate at relatively high speed in order to achieve optimum revolution rate required for the generator rotor. So the wind pump requires high torque and hence slow speed while the generator needs to turn fast with relatively low torque. High torque needs more blades while high speed would need fewer blades in accordance with wind disturbance concepts discussed above. Aerodynamic shapes of the blades may also vary in order to fulfil the different requirements. Another design consideration is the choice of the ratios of the coupling gears. For electricity generation, the gearbox system must be carefully designed to efficiently transfer the aerodynamic torque from the rotor blades to the electric generator. Finally the decision has to be made whether the machine should be a vertical axis wind turbine (VAWT) or a horizontal axis one (HAWT). These are the technically acceptable orientations of the axes of rotation of the blades. The advantage of the vertical axis turbine is that it does not require any mechanism to align the blades with the wind direction. They operate at all possible wind directions. The structure of the tower must be able to support the wind turbine and its components at a safe height from the ground. Although wind energy is renewable, it has some advantages and disadvantages. The advantages, in addition to it being a renewable source are:

- It is free and environmentally benign, producing no emissions or chemical wastes.
- Reduces dependence on fossil fuels.
- Its energy can be stored in batteries or as potential energy of pumped water.
- The technology is already well-developed with available equipment.
- Does not restrict land-use where it is installed.
- Adequate wind for power generation is fairly well-distributed around the world.

Some of the disadvantages are:

- Wind energy is intermittent.
- Wind machines can be noisy and an eye-sore to some people.
- Difficulty of repairing a faulty machine on a high tower.
- Most machines are very expensive for small operators.

5.5 Small Hydro Energy

The technology of converting kinetic energy of water in other forms of energy is a well-established procedure and has been used to generate power at competitive rates for centuries. As a result of this, it has become one of the major sources of power for many countries especially the less developed countries in Africa, Asia and South America. The technology was originally developed on a small scale to serve communities in the vicinity of the generator but as knowledge base expanded it became possible to generate power in a large scale and transmit it to distant locations. Large-scale hydropower generators make use of expansive water reservoirs achieved by constructing special dams to hold the water, and in order to accomplish this, large land area is usually required. Therefore there has been a growing concern about the impact of such developments on the environment and ecological systems. These considerations and the high cost of transmission have revived interest in small-scale hydropower production. Initially, during the early stages of the development of this technology, generation of electricity was not the main objective. Waterpower was mainly used to produce mechanical work to accomplish desired tasks such as water pumping (domestic supply and irrigation), grain grinding and machine operation for industrial activities.

Large scale centralized hydropower facilities have proved to be expensive and environmentally destructive with adverse destabilization of ecological systems. Experience has also shown that they are the cause of high transmission costs and hence high cost of electricity. Furthermore, there are very few rivers in East Africa that can continuously and reliably support such facilities. But there are small rivers that can be used to generate electricity on small-scale basis. These are the resources that should be effectively utilized to provide electricity to the scattered rural homes.

In addition to rivers, there are other ways of obtaining power from water resources. For example, ocean thermal energy, tidal power, wave energy and even geothermal energy are all water-based energy resources that can be utilized. With the exception of geothermal energy and hydropower, the use of all the other water-related energy resources have not made any significant impact on global energy supply systems. Even hydropower generation, which is one of the oldest technologies in electricity generation and is today well developed and exploited in a large scale, contributes only about 3% of the world's total electricity generation. The potential of hydropower as a source of energy in the continent of Africa is higher than that of Western Europe and is comparable to that of North America. The unfortunate situation is that Africa has the highest level of undeveloped hydropower potential in the world while millions of its people do not have access to electric power.

In order to appreciate hydropower technology, it is important to understand the principles behind water energy conversion to other forms of energy. The principle of hydropower utilization involves converting the potential energy of water in the reservoir into a free-fall kinetic energy, which can be used to do mechanical work. This means that the water storage facility must be at a higher altitude than the energy conversion point, e.g., electricity generator. The volume and direction of free-flow is controlled by using pipes, which direct the water to the point where the conversion is to take place. The power, P_o, in the water therefore depends on the vertical height, h, through which the water falls and the volume flow rate, V_f, of the water through the pipes. The speed of rotation of the turbine wheel is also crucial as it determines how much power can be generated. This rotational speed depends also on other parameters such as the diameter of the wheel. We begin with power of the flowing water, which can be expressed as:

$$P_o = \rho g h V_f, \tag{5.43}$$

where ρ is the density of water, and g is the gravitational force. Since the water is flowing inside the pipes, there is friction which reduces the effectiveness of the vertical height so that, finally, the power obtained from the water is as if it fell through a shorter height, h_e. This effective (or available) height is obtained by subtracting friction correction height, h_f, from the actual vertical height, h, i.e.,

$$h_e = h - h_f. \tag{5.44}$$

The magnitude of the friction correction height increases with the length of the pipe and therefore it is important to keep it at a minimal level by using short pipes. This means that the slope for water flow must be very steep. The hydropower generating assembly shown in Fig. 5.16 is over simplified. In reality, the cups or buckets on the turbine are many and quite close to each other particularly for the Pelton type turbine, which is one of the most common turbine designs. Moreover, the ends of the conveyor pipes are modified into water jets that direct water straight into the cups or buckets. This ensures that very little water power is wasted. The change in momentum of the water is therefore associated with the velocity of water from the jet, u_j, and the turbine velocity, u_t. We recall that change in momentum per unit time is the force that water applies on the turbine

$$F_T = 2\rho V_f (u_j - u_t), \tag{5.45}$$

so that the power extracted from water is

$$P(u_t) = F_T u_t = 2\rho V_f (u_j - u_t) u_t. \tag{5.46}$$

This power is maximum when the gradient, $dP(u_t)/du_t = 0$, giving $u_t/u_j = 0.5$. If all the kinetic energy of the water is extracted so that, behind the turbine, water falls vertically due to gravity only, then the output power of the turbine should be

$$P_j = \frac{1}{2} \rho V_f u_j^2, \tag{5.47}$$

where P_j is the total kinetic energy leaving the jet per second, i.e., power from the jet. In this case, the efficiency, h, of this turbine is 100%. However, because of other factors, the efficiency of a real turbine cannot be 100%, but normally ranges from 50 to 90%. The actual mechanical power from the turbine is

$$P_{mT} = \eta P_j = \frac{1}{2}\eta\rho V_f u_j^2 = \frac{1}{2}\eta\rho A(2gh_e)^{3/2}, \qquad (5.48)$$

where

$$u_j^2 = 2gh_e. \qquad (5.49)$$

The area, A, is determined to some extent by the size of the cup or bucket so that all the water from the jet falls into the cup. Nevertheless, the total flow in the turbine is always less than the flow in the stream. The total turbine sizing is based on the power generation output required from the system. The rotor of the turbine is connected to the rotor of the electric power generator through gear couplings or pulley linkage, so power generated will depend on the angular velocity of the wheel. If the wheel has radius R and turns at an angular velocity ω, then the power would be

$$P_{mT} = F_T R\omega. \qquad (5.50)$$

It is clear from (5.50) that the larger the wheel, the lower the angular velocity since $\omega = u_t/R$. At maximum power conversion, $u_t = 0.5u_j$ and this makes $\omega = 0.5u_j/R$, hence

$$R = 0.5(2gh_e)^0.5/\omega \qquad (5.51)$$

If the nozzle radius is R_j, then the nozzle area, $A_j = \pi R_j^2$, so the nozzle radius can be determined from the following relationship:

$$R_j^2 = \frac{P_{mT}}{\eta\pi\rho n_j\sqrt{2}(gh_e)^{3/2}}, \qquad (5.52)$$

where n_j is the number of nozzles, which is usually between two and four. From (5.52), the ratio of R_j to R can be determined

$$\frac{R_j}{R} = 0.68\varphi(\eta n_j)^{-1/2}, \qquad (5.53)$$

and the factor φ is known as the shape factor, and determines the best operating conditions for the system. Its important variables are the angular velocity, effective vertical height and maximum turbine power.

$$\varphi = \frac{P_{mT}^{0.5}\omega}{\rho^{0.5}(gh_e)^{1.25}} \tag{5.54}$$

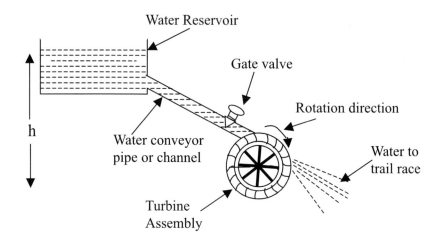

Fig. 5.16. A simplified hydro-power generator system

Figure 5.16 shows the major hydro components of a simplified method of converting water energy into mechanical energy of the turbine. There are three possible locations for the turbine assembly. If water from the channel is directed to the top of the turbine wheel as in Fig. 5.16, then this is the overshot water wheel type. The second type would be the case where water is directed to the middle part of the turbine wheel and this is the breastshot type. The last case is where water flows to the bottom of the turbine and this is the undershot water wheel type.

As discussed in Chap. 3, undeveloped hydropower potential in Africa is high and its future exploitation should put emphasis on small-scale hydro-generating facilities that have minimum impact on the environment and existing eco-system. Small and micro hydro schemes are cheaper and more viable for the African conditions where long distances between randomly scattered settlements and difficult terrains make it too expensive to distribute power from a central generating facility. Modern hydro turbines, if properly designed, are highly efficient with energy conversion efficiencies of over 90%. Two types of turbines have been in use for many years. These are impulse and reaction turbines and the size of the hydropower facility dictates which type would be most suitable. However small hydro technology, which should be of special interest to East Africa in particular and to the developing countries in general, is not a replica of large hydro technology even though the basic principles as discussed above are similar. Small hydropower systems should have simple and cheap layout techniques that do not require highly sophisticated and specialized knowledge or special equipment. They should require minimal civil construction work. In order to apply these requirements of simplicity and low construction cost, one must also accept lower efficiencies compared to the technologies, which are direct replicas of large-scale hydros. The combination of these aspects is one way of reducing investment costs per unit power since the larger part of hydro investment is made up of the cost of equipment, construction and special services and these form the bigger part of hydro energy price. Further investment saving can also be gained by not paying too much attention to reliability of the delivered power as in the case of large hydro facilities. Depending on the nature of the demand, water shortages for a few days in a year should be acceptable as this could save on civil works such as inlet structure. Small hydro schemes have additional advantages in that the power on the axis of a turbine can be made useful as a motor to directly drive apparatus such as food processing equipment, rice hullers, grain mills, oil press and sugarcane crusher. All these applications require very flexible rotational speeds and therefore are not too sensitive to fluctuation of water flow rates. The restriction, however, is that the generated power has to be used close to the site itself.

For these small-scale applications of small hydro systems, technologies are available for simple load controls and even use of free stream turbines, which do not require special built-up pressure heads. A more common example of such turbines is the undershot wheel placed in a river. There are also new technologies that simply use the wind energy conversion principles in designing hydro turbines in which aerodynamic shapes of the blades are used but since water is denser than air, there are some technical limitations to the practical use of such turbines. Load control is another construction design that can increase the cost of the facility. Ideally production of power and demand should be equal at all times even if no storage is provided. For small hydropower schemes, demand should be regarded as a given value, and then production is regulated as a function of the demand by closing or opening the inlet valves. This requires reliable and expensive valves and controls and, in some cases, surge chambers. To avoid this, it may be necessary to regulate the demand rather than production because doing this is cheaper, and more reliable control can be achieved. Demand can be divided into priority classes and then an electronic governor is used to switch production to these priorities according to their ratings so that the lowest ratings are provided with power only if there is enough power to cover them. Such a load governor is more effective and less labour intensive than the mechanical production controls, and they are cheap and readily available.

The following are some of the design considerations for small hydropower conversion systems. Energy in the water can be determined as:

$$E = gh + \frac{P}{\rho} + \frac{1}{2}v^2 (JKg^{-1}), \tag{5.55}$$

where h is the original water level above the turbine and v is the speed of the water. Since the water entering the turbine should have a higher energy than the water leaving the turbine, the net head on which power rating for the turbine should be based can be determined as:

$$H_n = \left(h_1 + \frac{P_1}{g\rho} + \frac{v_1^2}{g} \right) + \left(h_2 + \frac{P_2}{g\rho} + \frac{v_2^2}{g} \right). \tag{5.56}$$

Subscripts one and two refer to the water entry and exit points at the turbine respectively. The speed of the water v can be expressed in terms of the volume flow rate Q so that $v_1 = Q/A_1$ and $v_2 = Q/A_2$. Power to the turbine can then be expressed in terms of Q and H.

$$P_d = 9.81QH (kW), \tag{5.57}$$

and output power of the turbine is

$$P_t = \eta_t P_d, \tag{5.58}$$

where η_t is the efficiency of the hydraulic turbine.

There is no strict quantitative definition of small hydro systems but generally those that are rated up to about 10 MW are considered to be small hydro plants. Impulse turbines in which there is no change of potential energy of the water and energy transfer occurs between jets and runner blades are mostly used in small hydro systems. Some of these are the Pelton, Turgo and Banki-Mitchell turbines. Reaction turbines, which are normally submerged in water so that water fills the casing, are used in both large and small hydropower systems and the notable ones are Francis and Kaplan turbines. Both Francis and Turgo turbines are generally preferred for low-pressure heads of 15 to 40 meters but in general impulse turbines are used for relatively high-pressure heads, as high as 800 meters.

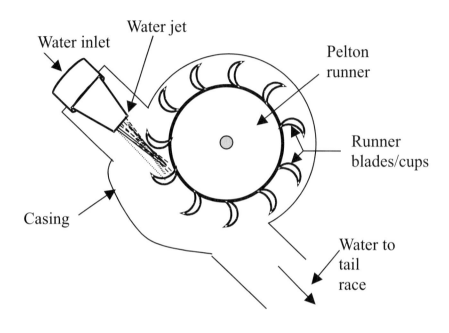

Fig. 5.17. Pelton impulse turbine

5.6 Biomass Technologies

Biomass is the oldest source of heat energy, which is obtained through combustion of organic materials such as trees, crop residues and other vegetation. The process that produces energy in plants is known as photosynthesis, which takes place only in the presence of sunlight. Biomass energy can therefore be considered as solar energy stored in the form of chemical energy in plants. Technologies have also been developed to convert raw biomass material into other forms of fuel. For example, biomass has been used for many years to generate electricity and to

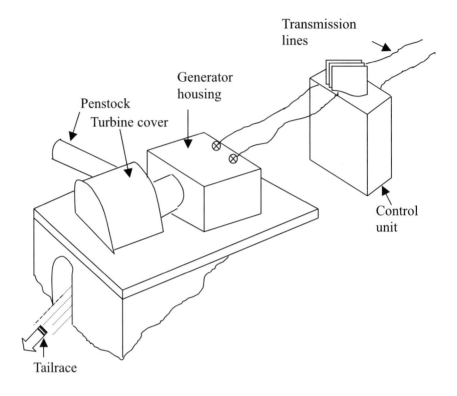

Fig. 5.18. A simplified small hydroelectric generation system

produce mechanical power via steam generation. Through a process known as gasification, biomass can be converted directly into producer gas, which can be used to power heat engines used for various purposes including electricity generation. Biomass is therefore a very important primary source of energy for many applications. Globally, biomass contributes between 13 to 16% of the total energy consumption but in the developing countries this contribution is much higher. For example, in East Africa it is about 85% on average. Although in East Africa biomass is used in its traditional form as firewood or fuel wood, in general, it can be transformed into many different forms as discussed above. The most common and simple technology is the conversion of woody biomass into charcoal, which is bunt in special stoves to produce the required heat for cooking and other heating requirements. This technique is widely practiced in many developing countries. Another rather indirect source of biomass energy is the use of dung from ruminant animals that feed on biomass materials. The dry dung burns slowly and is a source of heat energy used particularly by nomadic pastoral groups in the developing countries. Such people are still found in many parts of East Africa and are still burning cow dung to produce heat energy for cooking and space heating. Other conversion processes lead to the production of ethanol, biogas and other combustible gases. Small granules of biomass materials such as sawdust, coffee and rice husks and small leaves that usually burn too fast and hence unsuitable for cooking, can be briquetted into more compact forms that can burn slowly just like wood or charcoal. The chemical compositions of biomass materials vary from plant to plant and therefore not all types of biomass materials can be efficiently converted into a desired form of energy source. For example, plants like sugar cane produce chemical compounds than can be efficiently converted into ethanol. Another plant may not produce ethanol, at least not in significant quantities, as sugar cane would do. Similarly, wet cow dung is good in producing biogas (mainly methane) because of the fact that it has been somehow processed in the cow's stomach and so provides suitable environment for the action of microorganisms that produce the gas. All these conversion options and the wide range of applications of biomass as an energy source embrace all levels of technologies from very simple traditional ones involving direct combustion to highly sophisticated modern technologies like ethanol production, gasification, steam engines and co-generation of electricity. In this section, we will focus on the aspects of biomass technologies that have been practiced by the poor majority in East Africa. This will enable us to understand why it is important to improve on these technologies while searching for viable replacement for biomass energy. The improvements, which should be considered as short term measures will not only protect the environment but will also protect the users from health hazards associated with the use of biomass energy. However, the time each country will take to significantly move away from the use of biomass will depend on the rate of poverty reduction in the rural areas.

5.6.1 Biomass: The Traditional Use

As has been noted, biomass is the traditional energy source in many developing countries and is generally used as a direct source of heat for cooking. Unlike other sources of energy, the use of biomass does not require high and expensive technology. Cooking, for example, is achieved by simply burning the material in a suitable cooking stove.

For many years and even today, there was no specially designed wood stove. The so-called three stone fireplaces (Fig. 5.19) were and are still the method of cooking with wood fuel in most rural households in East Africa.

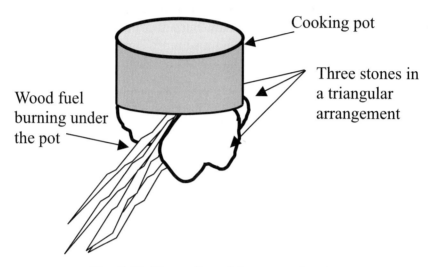

Fig. 5.19. The traditional three stone fire place

This method of cooking is the cheapest as it does not require any significant investment. But it also has many disadvantages. First, it is very inefficient in terms of heat utilization since the burning woodlots lose a lot of heat through the spaces between the stones. Second, the smoke from the burning woods is emitted in all directions so that the room is constantly filled with smoke during cooking. Third, they were suitable for round traditional clay pots, which are now not commonly used. Instead, metallic cooking pots with flat bottoms are today used on the stones and therefore the arrangement provides very unstable equilibrium for the pots. This results into frequent accidents and injuries to the users. Somehow the inefficient use of heat energy has been solved by using cow dung, wood ash or clay to cover the two spaces between the stones that are not used to fuel the fire. This significantly improves the efficiency but does not solve the problem of smoke; in fact it makes it worse. The problem of the unstable equilibrium of the cooking metallic pot has not been solved. Attempts to address these problems will be discussed later but for now, we look at another traditional cooking method that uses charcoal. Charcoal production is also an old traditional technology but to use charcoal, one needs a different stove-design because charcoal does not burn like wood. It compacts better than wood, leaving very little room for air that is required for combustion. A charcoal stove must use grates under the charcoal to allow air to flow through the charcoal pile to facilitate combustion. Traditionally, the metal stove shown in Fig. 5.20 was used for charcoal.

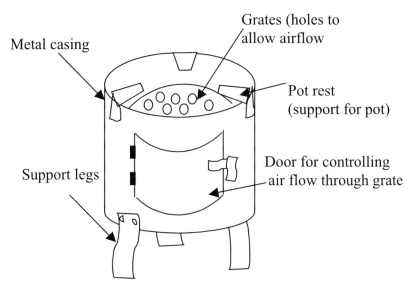

Fig. 5.20. Traditional Charcoal Stove

The metallic charcoal stove, like the traditional three stone system, is inefficient and becomes too hot when in operation. It is also not stable especially if a large pot is used. The door serves to control the rate of airflow through the burning charcoal and is also used to remove the charcoal ash. The stove is made of recycled metal and, due to the high temperatures, wears out very quickly. But it is cheap and a new one is usually easily acquired from the widely distributed vendors who are also able to repair the old stoves. There are a variety of shapes of the traditional charcoal stoves but the basic construction material and principle of operation remain the same throughout East Africa. The stove shown above is a typical Kenyan metal charcoal stove although there are also other shapes in some parts of Kenya. Since early 1980s, a lot of research work has been done to improve the efficiencies and the application conditions for both the three stone fireplace and the traditional charcoal stove. These improved stoves were promoted while at the same time recognizing the need to also encourage rural communities to plant trees in order to improve the supply of wood to meet the increased demand. Thus the 1980s saw a lot of efforts and donor funds directed to improving household cooking conditions and tree planting activities. This came at a time when the whole world was very concerned about the effects of global warming and rapid desertification process, which were closely linked to energy consumption. As a result of these concerns, the development of improved stoves and energy supply in general became some of the top priorities for international development agencies, particularly in the developing countries where biomass, an important source of fuel, was becoming scarce and eminent desertification process was threatening food security. The biomass situation was serious and had to be addressed. East Africa was one of the regions that benefited from this global concern and support from donors. Technologists and energy specialists, encouraged by this good will from the international communities started searching for socially acceptable efficient stoves. The major objective was to look for ways and means of reducing wood fuel consumption, which was believed to be associated with a number of environmental problems and, at the same time, reduce health hazards associated with the application of wood fuel as an energy source. Vigorous collaborative development and testing work were carried out in research institutions and private workshops until some satisfactory models were obtained. Finally, a few stove design proposals were presented as improvements of the three stone methods and also of traditional charcoal stove. Unfortunately these improvements were purely technical successes but the social aspects of their applications were ignored during their development and this became a major challenge when the new efficient stoves were to be disseminated. Wide-sale dissemination of the proposed stoves became more difficult than was envisaged and a lot of money was spent in this effort without much success.

In East Africa, Kenya took a leading role in the development and dissemination of the new improved stoves. The Kenyan experience later proved to be very useful to the Tanzanian and Ugandan efforts. However, the dissemination challenges arose from the fact that some economic aspects of the improved stoves, in addition to the social dimensions, were not considered in product development. For example, the new stoves could not accommodate the versatile use of fuel. Rural stoves often used various fuels ranging from cow dung to agricultural residues and, in many cases, the fuel is collected rather than bought. On the other hand urban stoves often used single fuel and the fuel is normally bought and not simply collected. The widely used stoves were also so cheap that the user could easily replace damaged ones. On the other hand, improved stoves were more efficient and therefore saved fuel and hence money and also provided the necessary stability and insulation that reduced accidental burns. The superior and more complete combustion significantly reduced the emission of toxic gases such as carbon monoxide. The serious handicap that caused a lot of concern was the fact that they were too expensive for the people to afford and, if poorly made, would not last long. Thus low quality and high cost were the most prominent disadvantages of the improved stoves that forced the scientists and energy specialists to go back to the drawing board. In the meantime, community-based groups also developed some interest in finding the solutions to these problems and gave the experts another dimension of collaboration. It turned out that this new collaboration facilitated the participation of the communities in the stove development process and finally made it possible to produce acceptable stove designs. In Kenya, this process gave rice to the production of the largely successful stove known as the Kenya Ceramic Jiko (KCJ), Kuni Mbili and the Maendeleo stoves (wood fuel stove). Whereas the latter two use firewood, the KCJ uses charcoal and is basically an urban stove that has permeated into many parts of the rural areas. It is made of metal cladding with symmetrically wide base and top (Fig. 5.21). The top part is the fire chamber that is fitted with specially designed ceramic liner to contain the heat. The bottom of the ceramic liner is perforated (the grate) to allow air to flow through charcoal for improved burning. One of the problems that affected the durability of the stove was lack of suitable material to strongly bind the ceramic liner to the metallic cladding. The damage of the clay liner and its separation from the cladding was the common fault that prohibited earlier attempt to disseminate the stove. Eventually tests done in research laboratories indicated that a mixture of vermiculite and cement or diatomite and cement could bind the metal and the clay liner well enough to withstand the cycles of high temperatures. These are the materials that are still used today for a durable KCJ. The stove has three strong triangular pot rests which can be rotated outwards to allow insertion of a new liner, if necessary, a feature that makes it possible to repair the stove. The door, located in the lower chamber below the grate, is used to control airflow through the charcoal and also for removing the charcoal ash. The door is opened during lighting but once the charcoal starts burning, it can be shut to facilitate slow burning process or opened for fast burning, whichever is desired.

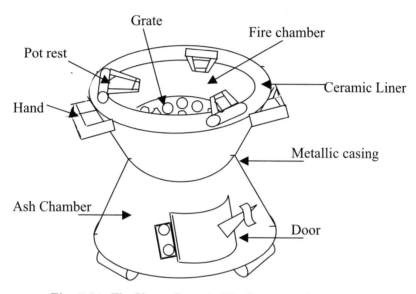

Fig. 5.21. The Kenya Ceramic Jiko (Improved Charcoal stove)

The KCJ was a close resemblance of the traditional metal stove but with added efficiency of up to 40% above the traditional stove and improved safety measures such as increased stability and reduced outside temperature of the stove. Initially, three sizes (small, medium and large) of the KCJ were recommended but, later, it became necessary to up-scale the size with appropriate shape modification for use by institutions such as schools and hospitals. The stove was successfully disseminated and the technology transferred to the local artisans. Its commercial production is today a mature cottage industry that is well organized with reliable sources of various components. Women who are the traditional specialists in pottery are producing the ceramic liners while artisans in the informal sector produce the metal casing. The assembly of the whole stove can be done by any of these groups or a third independent enterprise. Some more sophisticated manufacturers have mechanized their production systems while others use semi-mechanized procedures. All these production techniques, from manual to full mechanization, and the competition that settled in, have greatly reduced the cost of the stove. It is estimated that 800,000 KCJ are in use in Kenya and the number is still expected to rise. Some of these find their way into Tanzania, Uganda and as far as Ethiopia, Rwanda and Burundi. The experience, success and subsequent specialization of the production line have been replicated in both Tanzania and Uganda but the level of success is still well below the Kenyan case. In some arts of Tanzania like Dar-es-Salaam, the improved stove programmes started by importing KCJ from Kenya and testing their popularity. The stove would then be modified in response to the comments received from the users while at the same time training technicians to make them. This approach worked well in Dar-es-Salaam, Tanzania, where modifications to the KCJ were made to suit existing Tanzanian production technologies, giving rise to a new stove known as Jiko Bora which is the most widely used stove in Tanzania with an annual production of more than 500,000. Other stove designs were tried in various parts of Tanzania but due to low durability and technical problems, they did not gain any popularity as much as the Jiko Bora. The Morogoro clay charcoal stove is one example of a stove that was too delicate to withstand the cooking habits of the users. In Uganda, in addition to the use of KCJ, two improved stoves, which also are basically KCJ adaptations, differing only in shape, emerged: the Usika Charcoal Stove, and the Black Power Stove. But they both could not be disseminated at the expected rate by the producers. In general, all the new improved stoves were 3 to 5 times more expensive that the traditional metal stoves but as the price of charcoal increased, more and more people began to appreciate the need to use improved stoves. The types of stoves and even the names may differ in various parts of East Africa but the bottom line is the same in that they are all improved charcoal or wood fuel stoves whose operations are based on the same principles. The manufacturing process that involves division of labour is also a common characteristic in East Africa that has attracted many young men and women into stove-related trades. The amazing aspect of these stoves is that they all went against the dictates of economics to become more popular than the cheap traditional metal stoves despite the fact that the producers do not carry out any promotional activities but depend on customers who go to them. Taking the KCJ as an example, its efficiency was the immediate advantage that was recognized by the pioneer users and the word quickly spread round. Its charcoal consumption was more than 50% lower than that of the traditional metal stove and so the savings on charcoal was substantial. It was also more stable and produced less smoke and therefore was more suitable to the living conditions of the middle and low-income urban groups. These features appeared to be more important to the users and quickly pushed up the demand so that the cost became a secondary issue. Its popularity in urban areas then slowly diffused into the rural areas. The introduction and use of these improved charcoal stoves may be considered a success not just in East Africa but also in the neighboring states such as Rwanda, Burundi, Ethiopia, Sudan, Malawi and Somalia where KCJ in particular or its modifications are in use. However, the future of this development is not all that secure. Too many people are today involved in the manufacture of various shapes and sizes of these stoves that it is not easy to monitor the quality of the stoves. Consequently there are a number of low quality improved stoves in the market. Use of poor clay liners, low quality metal claddings and cheap bonding materials are increasingly becoming common practice. This and the high price will eventually encourage the users to revert back to the cheap traditional metal stoves, which are still very much available in the market. Table 5.1 below gives the number of improved stoves in East Africa.

Table 5.1. Number of improved charcoal stoves in East Africa (1995)

	Urban	Rural	Total
Kenya	600,000	180,000	780,000
Uganda	52,000	n.a	52,000
Tanzania	54,000	n.a	54,000

As has been mentioned the discussed improved stoves were basically urban devices and this is clear from the figures given in Table 5.1. There were also efforts to also improve the three stone cooking arrangement for rural application. However in terms of international support and the involvement of local non-governmental organizations, more interest was directed to charcoal stoves because charcoal production was considered to consume more wood resources than direct burning. There is no reliable data to verify this notion but this interest could be attributed to the commercial nature of charcoal as an energy source and the potential commercial gain that were expected from the new improved stove. The improvement of rural three stone fire place would not have any economic advantage and for this reason much of the improvement work was left to women groups and other community-based organizations. This is how Maendeleo stove got its name because it was developed through coordination of Maendeleo Ya Wanawake Organization, which is a very powerful women's organization in Kenya. The major challenge in the development of improved wood stoves for rural application was the fact that rural households buy neither stoves nor fuel and therefore any new device had to be provided almost for free, otherwise it would not appeal to the rural communities. At that time firewood was readily available and convincing people to save it was going to be a very difficult task. Despite these constraints, initiatives to improve the performance of woodstoves made significant progress. In Kenya the result was the Maendeleo Stove shown in Fig. 5.22.

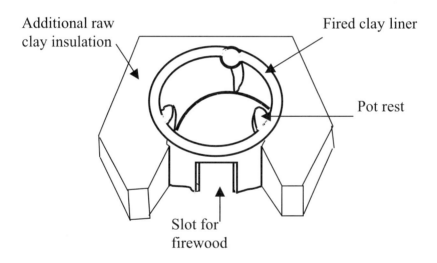

Fig. 5.22. Maendeleo wood stove

Maendeleo stove had to be cheap and had to resemble the three stone fireplace. It also had to be simple to construct so that women could install it themselves as they did with the preparation of the three stone cooking place. Thus the only new feature was to facilitate wood savings and reduction of emission of toxic gases by improving the combustion of wood. Several designs were tested but finally the choice was the Maendeleo stove. Thereafter, the developers embarked on a comprehensive training programme for women to acquire skills to make clay liners that would be used to replace the three stones. There was no metal component of the stove and therefore the liner was the only part that the user could spend money on. Like the KCJ, it also had some attractive features that encouraged rural households to buy the clay liner. It improved the cooking environment by emitting less smoke due to more complete combustion of the fuel and it also saved on wood by as much as 50% compared with the three stone method. Other attractions included reduced cooking time, less attention, safety and cleaner hygienic working area. A portable version of the Maendeleo stove was also developed under the brand name "Kuni Mbili", which means 'two pieces of fire wood'. It can be used either inside or outside the house and so, on moonlight nights, savings can be made on fuel for lighting by cooking outdoors. Kenya took the lead in the development of these and other large institutional woodstoves some of which are now produced in several production centers. Both Kuni Mbili and Maendeleo stoves are produced by women groups who also provide instructions on how to install them. It is estimated that there are over 300,000 maendeleo stoves in Kenya. The liners were introduced into Uganda and Tanzania by non-governmental organizations that participated in their development in Kenya. Modifications to the Kenyan design were carried out in these countries and are in use in different parts of East Africa. One important thing to note is that more than 80% of the population of East Africa live in the rural areas where the dominant energy source is biomass mainly woodlots but the stove development in the region focused more on charcoal stoves which are predominantly urban stoves. In deed these charcoal stoves have very quickly evolved as commercial items and there are already a very large number of people in the informal sector engaged in the stove business in one way or the other. On the other hand, good wood stoves have been developed but their demand is not as high as the charcoal stoves. This in itself is a good indication of the economic disparities between the urban poor and the rural communities. The urban poor are economically much better off than the rural people. There is also a question of attitude and availability of alternatives. The urban poor have no alternative but to use stoves for which they can easily get the fuel within the residential areas and are keen to save their incomes while the rural people have options some of which cost them nothing and so the idea of making some savings may not be meaningful to them. It is therefore very difficult to convince the rural communities to buy stoves when they have not been spending anything on them. Efficiency, better working environment and other advantages are not enough to persuade them to invest in the new device. Most of the improved stoves are therefore considered as items for the rural elite.

Wood is not only a household energy source but it is also used by schools, hospitals, hotels and other institutions where cooking is done for a large number of people. Usually these establishments buy wood in bulk from recognized suppliers once or twice a week. Their wood consumption rate is therefore easy to measure and so some savings can be appreciated. This is another area that drew a lot of interest as efficient stoves were being developed. Both scientists and energy specialists went through many cycles of designing, constructing and testing the so-called institutional stoves before suitable final product could be achieved. The process proved to be very expensive because institutions had their specific stove sizes that suited the size of the cooking pot as dictated by the number of people for which the food was to be prepared. These requirements meant that different sizes of stoves had to be developed and, in some cases, new sizes of the cooking pots had to be recommended to match the stove size. Due to the generally large sizes of the stoves and the associated high cost of producing prototypes for testing, very few individuals were involved in their development. In Kenya, only well established non-governmental organizations were able to reasonably invest in institutional stoves. One of the organizations that played a key role in the development of institutional stoves in Kenya was the Bellerive Foundation, which was based in Nairobi City. It finally developed efficient stoves known as the Bellerive Institutional Stoves that cut down fuel costs to the institutions by more than 50%. This was a significant saving given the large sum of money that the institutions were spending on firewood per month. The stoves came in sizes ranging from about 12 litres to 200 litres costing between US dollars140 to 1400. This included installation, training of the users and maintenance contract. Within eight years, over 500 institutions had ordered the stoves. Rural Technology Enterprises (RTE) was another organization that successfully produced efficient institutional stoves that were highly efficient. The sizes ranged from about 25 litres to 300 litres at prices from US dollars 250 to 1000. Rural Technology Enterprises

was more active and successfully distributed its stoves to many institutions including hotels, restaurants and departments of Home Science
complete with instruction manuals. The Kenyans were again ahead of
their counterparts in Uganda and Tanzania in this field and so institutional stove production in these countries borrowed a lot from Kenya.
In the middle of the 1880s two producers, having studied the Kenyan
stoves, emerged in Uganda with modifications to suit their situations
and brand names such as Usika and Black Power institutional stoves.
There are now a good number of institutional stove producers in both
Kenya and Uganda. Similar development has taken place in Tanzania
where there are other versions of institutional stoves, which have been
tested and approved for dissemination. The Duma Institutional stove,
for example, is already in use in many parts of Tanzania. All these
stoves have many common features. They are either cylindrical or rect-
angular steel structures insulated with fired clay (small clay bricks or
special clay blocks of appropriate shapes) and fitted with chimneys to
duct the smoke out of the building. The cooking pot, usually aluminium
cylindrical pot, fits inside the upper chamber of the stove, making it
more efficient in heat retention and safe to use. Firewood is cut into
short pieces of about 30 cm and fed at the bottom on steel grate that
facilitates airflow and allows easy removal of wood ash. One typical example is shown in Fig. 5.23. Most institutional stoves are durable with
a lifespan of up to ten years and the cost of acquiring one can be recov-
ered from fuel savings within less than four years. The grate is made
of heavy cast iron that can withstand the high temperatures achieved
in the firebox and this is the most difficult component to make and
perhaps the most expensive part of the stove. The stove is completely
insulated and the cooking pot normally fits inside the top part of the
stove. This design ensures that the environment around the stove is not
too hot for the user.

In Kenya, a number of institutions and non-governmental organizations are engaged in commercial production of various sizes of these
institutional stoves. For example, Muranga College of Technology developed a large fuel saving stove made of steel. The inner cylinder,
which during cooking is subjected to very high temperatures, is made
of mild steel while the outside cylinder is made of stainless steel. The
space between these cylinders is insulated with clay bricks that help to
retain heat. The stove also has a chimney to ensure that the smoke is
ducted out of the kitchen so that the cooking area is smoke free. The

sizes of these types of stoves range from 50 to 300 litres and can make food for up to 500 people. Muranga College of Technology is in fact using their own-made stove to cook for the students in the college. Before the introduction of the stove the college was using about 21 tonnes of fuel wood in four months at a cost of about US dollars 100 for every 7 tonnes but now it uses 14 tonnes over the same period at the same cost - a saving of about US dollars 300 per year. The cost of the stove varies with size and ranges from about US dollars 1,000 to 1,500 and Muranga was able to recover the expenditure within four years from fuel savings. Apart from fuel savings these stoves help cook the food quickly, keep the food hot long after cooking and keep the kitchen clean and healthy. Some of them can be fitted with water heating systems, which allow one to cook and heat water at the same time with the same amount of fuel wood. The Muranga stove has been sold to many institutions in Kenya including Nyeri Provincial, New Nyanza Provincial and Kerugoya District hospitals. Basically most institutional stoves have similar designs and operating principles and therefore the Bellerive type shown in Figs. 5.23[2] and 5.24[3] is a good representative. The Muranga design is however more expensive because of the use of mild steel inside the stove.

In all stove cases mentioned above, the technical problem was to reduce heat losses from the fuel and also ensure that the operator is both safe and comfortable while using the stove. In Kenya the research work was done at the Appropriate Technology Centre at Kenyatta University in Nairobi and involved the testing of various insulating and binding materials that could be readily obtained locally. The final stove for household application had to be light enough with low outside surface temperatures that would allow the user to handle it without any risk of getting injured. So the sizing of the various components was based on heat transfer considerations of the insulating liner. Since the traditional metal stove is basically cylindrical in shape, the original idea was to simply add suitable insulation to line the inner surface of the fire chamber and so heat conduction through a hollow cylinder was assumed to be the ideal approximation. Steady state heat conduction was considered in order to simplify the calculations. The radial rate of heat flow through a cylinder can be estimated as:

---

[2]Figure not drawn to scale
[3]Figure not drawn to scale

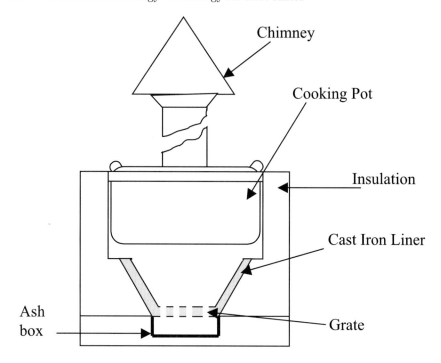

Fig. 5.23. The Bellerive Type Institutional Stove (Sectional View)

$$q(r) = -kA\frac{dT(r)}{dr} \qquad (5.59)$$

The solution of this equation gives the steady state rate of heat flow through the cylinder as:

$$q(r) = 2\pi kL\frac{(T_1 - T_2)}{\ln\left(\frac{r_2}{r_1}\right)}, \qquad (5.60)$$

where k is the thermal conductivity of the wall of the cylinder, L is the section length of the cylinder under consideration, r_1, T_1 and r_2, T_2 are respectively the inside and outside radii and temperatures respectively.

For the case of improved stove, the cylinder is made up of two materials: the insulator and the outside metal casing having different thicknesses and different values of thermal conductivities. It is assumed that the thin layer of the material that binds the insulator and the metal casing has the same thermal conductivity as the insulator and is therefore considered as part of the insulation. It is then assumed that if heat loss occurs only through radial flow then heat flux through the insula-

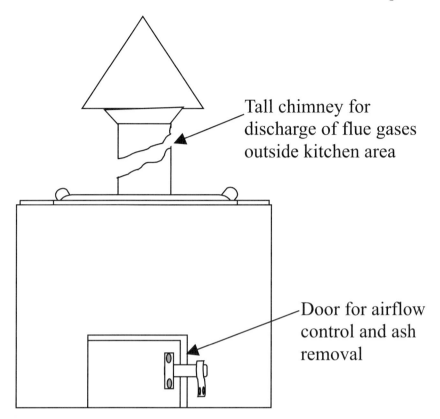

Tall chimney for discharge of flue gases outside kitchen area

Door for airflow control and ash removal

Fig. 5.24. The Bellerive Type Institutional Stove (Front View)

tion must be equal to heat flux through the metal casing. In this case the solution to the basic heat flow equation becomes:

$$q(r) = \frac{2\pi L k_1 k_2}{k_1 \ln\left(\frac{r_3}{r_2}\right) + k_2 \ln\left(\frac{r_2}{r_1}\right)} (T_1 - T_2), \qquad (5.61)$$

Where r_1 is the inside radius of the fire chamber, r_3 is the outside radius of the insulating material and r_2 is the radius of the metal casing. T_1 and T_2 are the temperatures of inside and outside surfaces while k_1 and k_2 are the thermal conductivities of the insulator and metal casing respectively. If the quantity of heat generated by the fuel under steady state conditions and thermal conductivities of the insulator and metal casing are known, comfortable outside surface temperature can be chosen so that the equation is satisfied. This makes it possible to determine the thickness of the insulation $(r_3 - r_1)$. The value of r_1 is normally chosen as the desired size of the fire chamber and the thickness of the metal casing is also generally known as the material is chosen, leaving only r_3 as the unknown radius that depends on the thermal conductivity of the material used as the insulator. Similar analytical procedure was used in designing other fuel-efficient stoves including Bellerive and Maendeleo stoves. Various materials were laboratory tested before suitable insulating and binding materials were chosen.

5.6.2 Biomass: Charcoal Production

The foregoing discussion was based on two forms of biomass fuels: wood fuel and charcoal. Wood is obtained from free access forests or is collected, with permission, from government-protected forests. Individuals or institutions that regularly consume a large amount of firewood grow their own trees for energy but these are usually not enough o sustain the consumption rate and therefore external sourcing still becomes necessary. A few people who know how to convert wood into charcoal, on the other hand, produce charcoal, and so charcoal is considered as a commercial commodity, which can only be obtained from the special producers. The art of converting wood into charcoal has been practiced for hundreds of years around the world particularly those areas where there is no coal. Charcoal itself has been used not only as fuel for powering steam engines but also as water purification medium. Other socio-cultural importance of charcoal has also been reported in some parts of the world. Thus charcoal is considered to be a more special form of biomass resource. The traditional practice where wood for charcoal is obtained free of charge from the nearby forests has also encouraged its production. Like in the case for firewood, people do not have to cut the whole tree to make charcoal but tree branches can be used. What is Charcoal and why must wood be converted into charcoal and then be used to produce heat energy which could be directly obtained from wood in the first place? The answer to this question will shed some light into the versatility of biomass as an energy source. Biomass, especially wood, is usually bulky and cumbersome to transport for long distances in large quantities and hence the need for conversion into other easily portable forms with higher heat content than raw wood. Such a form can be gaseous, liquid or carbonized solid. Charcoal is one of these forms in which heat is applied to raw wood in the absence of oxygen so that the wood is charred into a black solid substance, a process, which is normally accomplished by burning wood inside an airtight enclosure.

Traditionally, wood can be placed in a pit dug in the ground and then, after lighting the fire, immediately cover the whole pile with a layer of grass and thick soil to suppress direct combustion of wood. Charcoal, which is the final product, has high carbon content and is almost completely dry with no moisture content. These properties give charcoal more than twice heat content of dry raw wood and enables it to burn without smoke. That is, it simply glows red-hot allowing it to release heat slowly over a longer period of time than equal amount of wood. The characteristics of charcoal make it a suitable cheap energy for low-income urban dwellers that normally live in small crowded shelters. One disadvantage however, is that it produces relatively high level of toxic gases such as carbon monoxide which can cause death if not vented out of the room. Charcoal can be produced from many different types of biomass materials including agricultural wastes. Wood however is one of the most suitable feedstock in the form of roots, stems, and branches. Soft wood produces soft charcoal that burns quickly while hard wood produces hard charcoal that burns at a slower rate. Residues from timber factories including off-cuts can also be used to produce charcoal. The carbonization process that produces charcoal has four stages: controlled combustion, dehydration, the exothermic and cooling stages. Every piece of wood or any other biomass material must go through all these stages in order to become charcoal. During controlled combustion, one section of the biomass stack is burned using a large amount of oxygen so that the temperature of the rest of the stack is slowly raised to between 400 to 600°C. After attaining the high temperature, oxygen is cut off and the temperature decreases but dehydration process continues without combustion until all the moisture is driven out in the form of steam. Once the dehydration process is completed, the biomass material begins to crack and break down due to heat, releasing even more heat. This is the exothermic process during which no combustion should take place and so oxygen is totally cut off but the temperature of the stack increases to its highest level of about 600°C or more. After this natural cooling takes place and the end product is charcoal. The East African traditional method of making charcoal is to use a shallow pit or just a stack of wood covered with soil and here the person uses instinct rather than scientific knowledge in achieving the four stages of the process. The shallow pit or an anthill-like structure acts like the charcoal firing kiln. It is a simple earth kiln composed of stacked pile of wood completely covered with a layer green vegetation and earth.

Although this is the cheapest way to make charcoal, it is also very inefficient and clumsy. So recognizing that charcoal production is an activity that is there to stay, many attempts have been made to design and produce improved charcoal kilns. There are a variety of kilns but the operating principles are basically similar to that of the traditional one described above and some of them incorporate only slight modification from the traditional one. For example, the Casamance kiln, which is one of the improved versions is just a simple traditional earth kiln fitted with a chimney to create a more sophisticated gas-heat circulation modes that facilitates better control of the system and achieve higher conversion efficiencies. Modifications to the traditional earth pit kiln were also made in order to improve conversion efficiency and also shorten the carbonization time. The main modification was the introduction of a metallic lid to replace the layer of earth as a cover so that better airtight seal is achieved. Other improved kilns, which are not in use in East Africa, are the giant Missouri kiln and versions of Mark V kiln made of cement and metal respectively and have chimneys, more than one lighting points and appropriate number of air control holes. They take long to accomplish the carbonization process due to their large sizes and also have provisions for mechanized loading and collection.

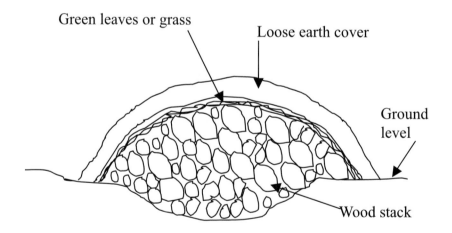

Fig. 5.25. Traditional earth kiln

In East Africa, mostly men produce charcoal in small quantities as an economic pastime activity, during the period when labour demand

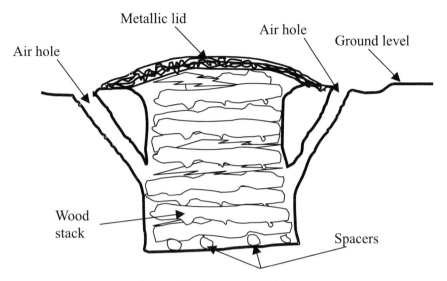

Fig. 5.26. Improved pit kiln

in other areas like agriculture is low. Many of these people do not see the need to invest in improved kilns since the activity is seasonal or circumstantial. For example, when a tree is felled and the user does not need the branches then these can be used to produce charcoal. Professional urban charcoal traders often do not produce charcoal themselves but go round in lorries buying charcoal from rural producers to go and sell in the towns. It is therefore a common site in certain parts of the country to see sacks of charcoal placed by the roadside waiting for any willing buyers. The introduction of improved charcoal kilns has not had any impact in East Africa partly due to the haphazard nature of charcoal production and partly due to the fact that good charcoal producers make very efficient earth kilns that are comparable to the improved ones.

5.6.3 Biomass: Briquette Production

There are many different types of biomass materials that are not suitable as sources of heat through direct combustion because they burn too fast with too much flame that causes discomfort to the user. Grass and dry leaves fall under this category. Others are too compact to burn and produce heat at the required rate. Sawdust and charcoal dust are in this category. These materials are usually considered as rubbish and therefore cause some environmental problems as they are carelessly disposed of within the vicinity of the residential areas. Occasionally the very poor people living in the slums of the towns collect and mould them into small hard round balls and use them as source of heat energy for cooking. The shape of these wastes allow them to be arranged in the stove such that there is just enough airflow through the pile to support combustion and the increased density also reduces the burning rate so that the right amount of heat is produced. This practice is a crude way of producing fuel briquettes. A more organized briquetting of biomass waste is considered as one way in which biomass fuel-base could be broadened to reduce shortages for the rural communities and urban poor because there are many biomass materials that could be used as energy but are not used due to reasons given above. These materials include rice husks, groundnut shells, bagasse, sawdust, coffee husks, straws of various cereal plants, grass and leaves that seasonally drop off plants. Briquetting these residues transform them into good quality fuels, which can be used in both wood and charcoal stoves. Although hand-pressing the material into a solid ball can produce some briquettes, it is better to use a simple machine that would produce high-density briquettes because the heat value of the briquettes increases with density. The use of briquettes as fuel has many advantages in that an otherwise waste material would be turned into a profitable resource and so reducing the environmental problems of disposing these materials into rivers and lakes. Secondly, the trees and other useful biomass materials will have more time to regenerate. An aspect that must be considered however is that agricultural residues are natural soil conditioners and their use as energy might negatively impact on the quality of agricultural soil. The other argument that has been put forward is that some agricultural residues are used as livestock feeds and that diversion of their use to energy would reduce their availability as animal feeds. These arguments apply only to a few biomass wastes because it has been observed that some of these materials are not in that high demand and that is why they are an environmental concern. Furthermore, production of briquettes is not an activity that everyone would want to engage in. The process, like charcoal production, is labour intensive as the waste must be collected, chopped, blended with suitable binding material and then compressed in some form of a press. The simple compressor is hand-operated (see, e.g., Fig. 5.27) and there are a variety of them with some using bolts and washers to squeeze the material into high-density pellets. The major task in the process is to find a suitable binding paste that can also burn without emitting toxic fumes. The briquette production rate is generally low because hand or foot operated machines can only produce a few pellets at a time, normally up to 50 kilograms per day. There are two ways of using the briquettes. The first one is to use it in a similar way as firewood: dry it after compression and then use it in a suitable stove. The second method is to convert it into charcoal and then use it in a normal charcoal stove the same way charcoal is used. These two methods have been practiced in East Africa for many years particularly in Kenya where small groups and even well established industrial manufacturers have been producing fuel briquettes for sale using either screw or piston press. The raw material input is usually a waste product from the same factory. For example, some paper manufacturers have been making briquettes out of sawdust and wood shavings while coffee processing industry in Kenya has been producing fuel briquettes from coffee husks. In Uganda, the Black Power group practiced small-scale biomass briquette production using coffee husks and sawdust but could not develop the business to maturity due to technical problems and poor economy of scale. Smaller women groups using portable parts of the manual compressor machine appeared to have made some progress in Uganda and successfully replicated the technique in other parts of the country. This apparent success was partly due to the use of old metallic tins and other available scrap materials to make some parts of the machine and largely due to the handy support received from local organizations such as YWCA and Small Scale Industries Association. A local binding material in the form of sticky staff from banana peels also came in handy, as banana is a common staple food in Uganda. One common problem with briquette technology is that production rate is generally so low that it does not justify the effort. It is however likely to succeed if produced in large quantities as an industrial commercial product for application at household level and as an industrial fuel substitute or supplement.

5.6.4 Biomass: Industrial Application

There are a number of industries using biomass materials in their main production lines and therefore produce biomass wastes that can be turned into useful products. Agro-based industries such as sugar and paper producing industries are good examples of such industries that have problems with disposal of biomass waste such as wood dust and

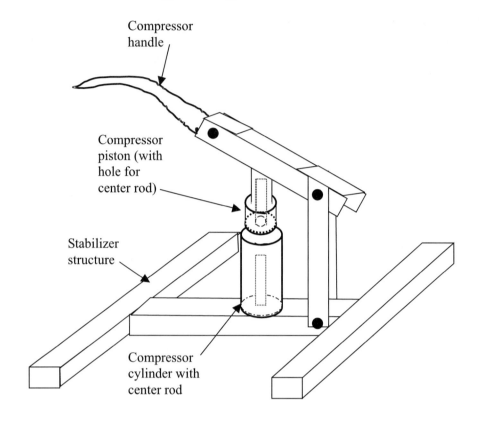

Fig. 5.27. A simple Hand-Operated Piston Type Briquette Machine

bagasse. There are two possibilities of making use of such materials. One way is to set up an industrial briquette production unit that would produce fuel pellets out of the waste and sell it as one of the industrial outputs. The second way is to use the material to produce electricity in a co-generation arrangement and sell it to the relevant grid power distributors. Both options will enable the companies to cheaply produce a second commodity for the local market and increase their profit margins. It has been noted that the market demand for fuel briquettes is still very low in East Africa but the national demand for electricity is high and so the industries are sure to gain from co-generation using their own biomass wastes. Paper Mills, sugar and Agrochemical Companies require steam in their normal industrial activities and co-generation, as the name implies, is combined heat and power generation, which these industries can accomplish with very little additional investment. This means that in addition to using steam for normal industrial application, they can also use that same steam to generate electricity, enabling the company to sell two products from the same process instead of one (see, e.g., Fig. 5.28).

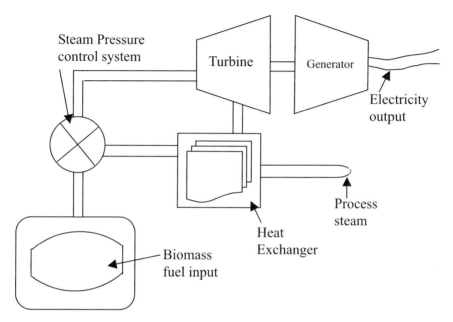

Fig. 5.28. Simplified co-generation system

Co-generation is based on steam engine principles, which is one of the oldest technologies for producing mechanical work from steam pressure. The biggest advantage is the fact that the raw material (bagasse, wood cut-offs, rice husks, coffee husks etc) is part of the normal waste that is produced as a result of the main activity of the industry. Co-generation, when considered from both economic and environmental points of view, has enormous advantages. It is an organized way of effectively using biomass waste, which is continually generated in agricultural production. Thus there is almost no possibility that the generation would run short of the fuel because both the farmers and the companies mutually depend on one another. Consequently, extensive co-generation, if implemented by rice, sugar and coffee millers, would lead not only to increased agricultural interest in these products but also to their outputs. It is also likely that the cost of electricity and even other products from co-generation systems would decrease. There are a number of co-generation systems that existing agro-based companies could consider adopting. Two of these are shown in Figs. 5.29 and 5.30.

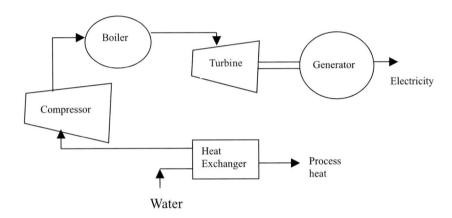

Fig. 5.29. Closed cycle non regenerative co-generation turbine

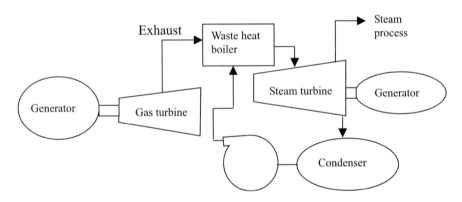

Fig. 5.30. Combined cycle co-generation system

As discussed in Chap. 3, there is a significant co-generation potential in East Africa, which is yet to be exploited and there are already indications that, with suitable sale agreements, the existing agro-based industries are willing to invest in commercial electricity generation. Support from national governments and cooperation from national electricity distribution companies would determine the extent of commercial co-generation practices in the region. There is, however, no doubt that co-generation has some of the best conditions for sustainable use of biomass energy resources. Many other industrial biomass energy applications such as tobacco curing, tea processing and brick-making are net energy consumers and therefore should actively support biomass regeneration activities, particularly agroforestry.

5.6.5 Biogas

There are two ways in which biomass materials can be converted into combustible gas. The common and simple technique is a process in which already processed biomass material can be further broken down by microorganisms and in the process produce methane - a combustible gas that can be used for cooking and lighting. The second one uses a more sophisticated method to directly convert biomass material into gas through a process commonly referred to as gasification. In this section, we will consider biogas production, which is a small-scale energy production activity that has attracted a lot of attention in East Africa, particularly in Tanzania where more than 1000 biogas digesters had been installed with significant support from the government. Both Kenya and Uganda had only about 500 and 10 respectively by mid 1990s. Biogas production technology is simple but the rate of gas production is generally slow and sensitive to weather conditions.

Biogas, which is sometimes referred to as bio-methane because its main constituent gas is Methane, CH_4, is a renewable fuel with properties similar to that of natural gas. It is produced from biomass materials in the form of solid or liquid waste such as cow dung. The production of the gas involves microbial action on the waste in the absence of air - a process known as "anaerobic digestion". The container in which the process takes place is called "digester". Digesters can have different shapes and sizes and may be used to accomplish some other necessary waste treatment processes. Conventional digester systems have been used for many years and from different wastes. One interesting digester is the type that is an integral part of sewage treatment plant to help stabilize the activated sludge and sewage solids while at the same time producing biogas for other applications.

Interest in biogas has led to the development of high-rate, high efficiency anaerobic digesters, which are also referred to as "retained biomass digesters" because they are based on the concept of retaining viable biomass by demobilizing the sludge. Digester types include batch, complete mix, continuous flow and covered waste treatment la-

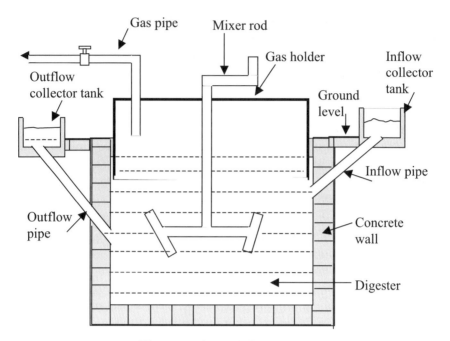

Fig. 5.31. A simple biogas digester

goon. Biogas technology by its very nature is a rural technology that can have significant impact on rural energy development strategies. There are many designs of digesters but the famous ones are the Floating Drum type developed and promoted in India and the Fixed Dome type, which was developed in China. The Indian type soon became obsolete as a result of the more durable Chinese Fixed Dome type, which can last from 20 to 50 years. There are still other types, for example the Bag Digester developed in Taiwan, Plug Flow Digester developed in South Africa and Covered Anaerobic Lagoon type used in the United States of America. The process of producing the biogas is however the same. In all cases the digester must provide anaerobic condition within it. The waste e.g. cow dung is first mixed with water usually at a ratio of one to one before pouring it into the inlet tank from where it flows into the digester (Fig. 5.31). If the input is too dilute, the solid particles will settle at the bottom of the tank and if it is too thick, it will impede the flow of the gas from inside the slurry. The first process inside the digester is the decomposition of plant or animal waste matter. During this process, the organic material is broken down to small pieces. Then the material is further decomposed and converted to organic acids, which are in turn converted to methane through some chemical reactions. The decomposition of organic matter and the subsequent release of the biogas is accomplished by some bacteria which survive well in anaerobic conditions. Thus the main component of a biogas plant is the anaerobic digester.

The bacteria that cause the release of the gas are very sensitive to temperature and pH changes inside the digester. Suitable digester temperatures for common species of these bacteria range from 30 to 40°C with an optimum at about 36°C. It is of course possible for the digester to operate outside these ranges but the system would be vulnerable if not closely monitored. After the digestion of the waste matter, the remaining fluid consisting of mostly water and stabilized waste can be used as manure. The actual digestion takes place in three main steps. The hydrolysis is the first step during which bacterial enzymes break down proteins, fats and compound sugars into simple sugars. The next step is the acid formation when bacteria convert the sugars to acetic acid, carbon dioxide and hydrogen. Finally the enzymes convert the acids to methane and carbon dioxide. Hydrogen, which is produced during acid formation stage, is then combined with carbon dioxide to form methane and water. The final biogas has traces of many gases such as hydrogen and nitrogen but the main constituents are methane (CH_4) and carbon dioxide (CO_2). An effective digester should have methane as the dominant gas followed by carbon dioxide such that the two constitute over 80% of the gas. An important factor that affects the operation of a biogas plant is the carbon/nitrogen ratio (C/N). A ratio between 20 and 30 is considered to be optimum for anaerobic digestion. It is important to consider this ratio when choosing the type of waste to use in the digester. For example, cow dung has a ratio of about 24, water hyacinth has 25, pig waste has 18, 10 for chicken waste and 8 for both human and duck wastes. Other materials such as rice straws and saw dust have much higher C/N ratios. If the ratio is higher than the optimum range then the little nitrogen will be consumed by the bacteria to meet their protein requirements and gas production rate will be reduced. On the other hand if the ratio is low, more nitrogen will be liberated and become available for ammonia formation and this will raise the pH of the system and hence reduce the population of the microorganisms. A pH higher than 8.5 is toxic to the bacteria and should not be allowed to develop inside the digester. There are many species of these microorganisms and their characteristics also vary but their sensitivity to the microclimate inside the digester is generally similar.

Although any biodegradable organic matter can be used as raw input into the digester, it is better to choose an input with high gas productivity. This is even more important if the material is to be purchased. However, care must be taken to ensure that the design of the digester is compatible with the type of input to be used. Cow dung which is a popular choice as an input material is easy to handle and can be obtained in large quantities in one location but its gas production potential ranges from 0.02 to 0.04 cubic meters of gas per kilogram and is just as good as that of human waste. Chicken waste has higher gas production potential than this, lying between 0.06 and 0.12 cubic meters of gas per kilogram while pig waste is between 0.04 and 0.06 m3/kg. Some plants such as water hyacinth have comparable gas production potential as cow dung. A mixture of these organic materials can be used especially if a material of low gas yield is more readily available than the high yield material. Similarly, materials with low Carbon-Nitrogen ratio can be mixed with those having high ratios so that compositions suitable for the bacteria are achieved. Another important factor is the amount of volatile solids in the slurry mixture. The higher the volatile solid content the higher the gas production because the gas comes mainly from these solids and not from the water content. In order to understand the chemical reactions that take place in the process of digestion, it is important to know the chemical composition of the waste input. These materials contain carbohydrates, lipids, proteins, and inorganic substances, which are first broken down to simpler sugars such as glucose with the help of enzymes released by the bacteria. The fine sugars are then further fermented into acids by enzymes produced by acid forming bacteria. The common acid is the acetic acid and some traces of propionic acid, butyric acid and ethanol. Finally, the following chemical reactions, catalysed by the bacteria, result into the production of biogas (methane).

$$CH_3COOH(\text{Acetic acid}) \longrightarrow CH_4(\text{Methane}) + CO_2(\text{Carbon dioxide})$$

$$2CH_3CH_2OH(\text{Ethanol}) + CO_2 \longrightarrow CH_4 + 2CH_3COOH$$

$$CO_2 + 4H_2 \ (\text{Hydrogen}) \longrightarrow CH_4 + 2H_2O(\text{water})$$

These reactions take place all the time, producing methane, which is accumulated as the main constituent of the biogas. In addition to controlling pH and temperature inside the digester, it is also important to pay attention to loading rate, retention time, and other materials that would increase toxicity. Over feeding or under feeding the digester will reduce the rate of gas production. The feeding rate would depend on the local conditions but in general, and for tropical conditions, feeding the digester with 5 to 8 kilograms of cow dung per cubic meter of the volume of the digester per day is recommended. The specific amount would also depend on the time required for the dung to stay inside the digester in order to be fully acted upon by the micro organisms. This is the retention time. For cow dung under tropical warm temperature conditions, the retention time is usually between 40 to 70 days but, in general, the higher the temperature, the lower the retention time. The performance of the digester can be significantly affected by the level of toxic elements inside the input material. Mineral ions, heavy metals and detergents are some of the toxic materials but small quantities of some of these (e.g., sodium, calcium, copper, zinc, lead and potassium) can stimulate the growth of the bacteria. To be on the safe side, recognizable amounts of these materials should not be allowed inside the digester.

5.6.6 Biomass: Gasification

Biomass solid materials can also be directly converted into combustible gas though a process known as gasification. The gas produced is often referred to as producer gas and can be used as a heat source usually to run a heat engine such as electricity generators or motor vehicle engines. The conversion process involves partial oxidation and pyrolysis in a gasifier unit. The conversion is possible due to the fact that biomass materials like firewood, agricultural residues and wastes are generally organic and contain carbon, hydrogen, and oxygen together with some moisture. Under controlled conditions of low oxygen supply and high temperatures, most biomass materials can be directly converted into a gaseous fuel. The thermo-chemical conversion of solid biomass into gaseous fuel is called biomass gasification. Although the gas has comparatively low calorific value, it can be burned efficiently using compact gaseous and liquid fuel-burning equipment (stove) with high thermal efficiency and good degree of control.

Gasification technology is more than a century old but interest in it diminished when other liquid fuels such as oil became more readily available. However, recent renewed interest in it because of increasing fuel prices and environmental concerns, has spurred its development to a fairly sophisticated level. Gasification process involves a series of complex thermo-chemical reactions which cannot realistically be split into strictly separate entities, but there are stages through which biomass must pass in order to produce the gas. The equipment used may look very simple and usually consist of cylindrical container with space for fuel, air inlet, gas exit and grate. The construction material can also be the normal building materials such as concrete, bricks, steel or oil barrels.

The main stages for the gasification process are Drying, Pyrolysis, Oxidation and Reduction. When the biomass material to be gasified is loaded into the gasifier, it is first dried at temperatures ranging from 120 to 160°C so that all the moisture is removed and converted into steam. At this stage the fuel does not undergo any decomposition. The dry biomass is then thermally decomposed in the absence of oxygen. This is the pyrolysis stage, which, at temperatures up to 700°C, produces three phases of the biomass: solid (mainly char or charcoal), liquid (acid or oil) and gas (mixture of mainly carbon monoxide and hydrogen). This is followed by oxidation stage during which air is introduced into the gasifier at temperatures in excess of 700°C that facilitates the reaction between the constituents of air and solid carbonized fuel to produce largely carbon monoxide and steam. Finally, the reduction stage, which also takes place at temperatures in excess of 700°C in the absence of oxygen, enables high temperature chemicals to react, producing more carbon monoxide, hydrogen, methane and steam. If complete gasification takes place, then all the carbon is burned or reduced to carbon monoxide and some other products, which are vaporized. But usually the remains are ash and char or charcoal. For the theoretical complete combustion of biomass, about 6 kilograms of air is needed for each kilogram of biomass fuel producing carbon dioxide and water as the main products. For pyrolysis however, only about 1.5 kilogram of air is required for every kilogram of biomass in order to produce the combustible gas, which is a mixture of carbon monoxide and hydrogen and traces of other gases in small quantities.

The raw producer gas contains tar and other particulate matter, which must be removed if the gas is to be used as fuel for engines.

Thus, apart from the four stages mentioned above, the gas must be "washed" to remove impurities and finally "dried" before use. If the final application requires only heat, then cleaning of raw gas is limited to the requirements of the thermal process. Although normally there are traces of many gases in the producer gas, a well designed gasifier should have a high percentage of the mixture of cabon monoxide and hydrogen, for example 55% CO and 25% H_2 . The reactions that take place inside the gasifier should be conditioned to produce the right mixture of the gases. Some of the reactions require heat input while others release heat. The following are a few examples.

$$C + O_2 \longrightarrow CO_2 + heat\ released$$

Oxygen in the air supports the burning of carbon and produces carbon dioxide and heat. At the same time, hydrogen in the air reacts with oxygen to produce steam.

$$H_2 + O_2 \longrightarrow H_2O + heat\ released$$

In the reduction zone, chemical reactions take place at high temperatures in the absence of oxygen so that the major participants in the reactions are carbon dioxide, carbon, hydrogen and steam.

$$C + CO_2 \longrightarrow 2CO + heat\ input$$

$$C + H_2O \longrightarrow CO + H_2 + heat\ input$$

$$CO_2 + H_2 \longrightarrow CO + H_2O + heat\ released$$

$$C + 2H_2 \longrightarrow CH_4 + heat\ released$$

As mentioned above, a good reactor should produce more of the combustible gases.

Gasifier sizes range from very large industrial complex to mini portable systems. The large systems are suitable for generation of electricity while the portable ones are used for running small engines like motor vehicle engines. It is not easy to classify gasifiers because there are wide variations in designs. If classified in terms of airflow, then there is updraft, downdraft, cross draft and twin fire gasifiers. The first three

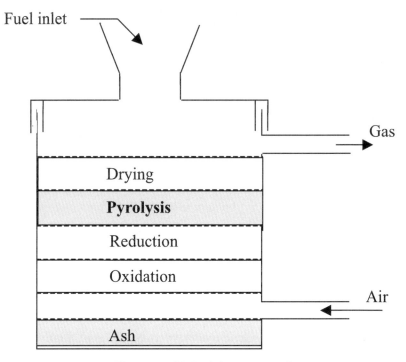

Fig. 5.32. Updraft biomass gasifier

are the most common types. Figure 5.32 is an example of updraft gasifier. They can also be grouped according to their operational modes, for example, screw auger reactor, pressurized solid bed, oxygen-blown and internal circulation fluidized-bed gasifiers.

In regions where biomass is already available at low prices or completely free like the East African rice mills, sugar factories and paper mills, gasifier systems offer definite economic advantages in addition to its environmental friendliness as a result of firewood savings and reduction in CO_2 emissions. The gas also has the potential to replace petroleum-based fuels in several applications and save valuable foreign exchange used in the importation of oil-based fuels. Further, from the gasification process, other products such as fertilizers, chemicals and plastics can be produced. It is also possible to obtain flame temperatures that are as high as $1200°C$ if the air is appropriately pre-heated and mixed with the gas making the gasifier suitable for heat treatment applications or as fuel for ceramic kilns and boilers. Like any other gaseous fuel, producer gas provides much better control of power levels than solid fuels and facilitates efficient and cleaner operation. When burned to give heat energy, thermal energy of about 4 to 5 MJ can be obtained from one cubic meter of the gas and flame temperatures of up to 1200°C can be obtained. It can also be used to blend fuels such as diesel in various ratios and still provide good power output.

Biomass is such an important source of energy in East Africa and needs detailed analysis in order to not only understand its full role but also be able to formulate a comprehensive energy policy that can address realistic energy planning strategies. It is therefore important to give detailed biomass energy analysis that can serve as an overview general case for East Africa.

5.7 Biomass Energy in Uganda: An East African Model

Biomass energy strategies in the three east African states are basically similar with only minor variations in the level of emphasis and implementation of some specific areas. We therefore consider in details the situation in Uganda as a representative case for the whole of East Africa.

5.7.1 Uganda's Biomass Energy Policy

In 1999 the Ministry of Energy and Mineral Development (MEMD) of Uganda began a process of formulating a new policy for the energy sector under the National Energy Policy. This energy policy is unique in that it includes a section on biomass energy. Revision of the Forest Policy began in 1998. Both the new and old policies (1988) attempt to address biomass issues. The Forest sector has gone further to develop a National Forest Plan, which is complementary to the National Environment Action Plan.

The National Biomass Study (NBS) of the Forest Department has collected data on biomass distribution throughout the country since 1989. This information is useful for planning for biomass supply and includes information on crop distribution and acreage and livestock stocking rates which can be used to establish how much agricultural residues are available for energy production. It should be noted that use of agricultural residues for energy production should not in any way compromise its priority use as a natural fertilizer since the average farmer in Uganda cannot afford commercial fertilisers.

The MEMD has developed strategies for rural electrification, petroleum supply and energy efficiency. Despite the importance of biomass in the national energy balance accounting for 93% of total energy consumption there is no comprehensive biomass energy demand strategy. Wood is the predominant source of energy for domestic cooking and for process heat in the industrial and institutional establishments. Charcoal production and consumption is an important part of the economy supplying most urban areas with cooking fuel and generating about 20,000 full time jobs. Due to the high rate of consumption, it is believed that biomass resources are being depleted at a higher rate than production but data are insufficient to prove this view. In addition, land under biomass production is regularly converted to farmland, which is sometimes devoid of trees. The MEMD and other actors have implemented isolated biomass projects.

This strategy attempts to synchronize the energy strategies in the energy policy, the national forest plan and the national environment plan. The proposed biomass energy strategy has been formulated on 6 major principles namely:

- Institutional aspects;
- Human resource development;
- Mobilizing financial resources;
- Dissemination of awareness;
- Quality control; and
- Research.

The 1995 Constitution of Uganda recognizes the need for an energy policy oriented to the poor when it states "The State shall promote and implement energy policies that will ensure that people's basic needs and those of environmental preservation are met". The importance of energy for the poor and the development of Uganda is recognized in the revised Poverty Eradication Action Plan (PEAP) of 2000. Energy has a direct impact on poverty alleviation. Improved electricity supply is expected to enhance poverty alleviation initiatives through both the promotion of private sector driven economic growth and via direct poverty impacts. The link between energy and the introduction of new technologies for agro-processing in the rural areas is especially relevant to government's Plan for Modernization in Agriculture. PEAP also highlights the link between energy and basic needs and notes that the dependence on fuelwood increases the burden on women and cause environmental degradation. It further notes that the first step in climbing the energy ladder is the use of improved cooking technologies and the introduction of more efficient methods for charcoal production.

About 93% of the energy consumed in Uganda is from biomass, which includes wood, charcoal and agricultural waste. 95% of wood supply is for energy consumption with the following characteristics:

- Wood is the only energy option for the poor since only about 1% of the rural population has access to electricity (2000 estimate).
- Scarcity of wood affects nutritional value of food cooked as it limits the type of food and level of cooking and therefore many people strive to ensure that it is available all the time.
- Wood is a renewable energy if its regeneration and consumption is planned sustainably.
- Wood is a major source of process heat in industries and is therefore of important economic value for the nation in addition to its direct use as construction material.

These characteristics confirm that wood will continue to be the dominant source of energy in Uganda for many years to come even if the entire hydro potential in Uganda is to be fully utilized. One major concern is that Ugandan households generally use biomass energy inefficiently. The application of improved wood stoves and other energy sources (LPG, solar energy, kerosene and electricity) is limited in most areas. A few institutions such as schools have converted from open fires to improved cooking stoves. A few high-income households and expatriates us LPG and electricity for cooking.

In Kampala the market efficiently supplies the consumer with a steady, uninterrupted supplies of fuel wood and charcoal at relatively low prices. However, these low prices are below their economic cost, due to the fact that fuel wood taxes and fiscal compliance rate are low and that so much charcoal is coming into urban areas as a result of rapidly increasing land clearing for grazing and agriculture. This is also confirmed in Mbarara Town where the price of the charcoal bag has reduced from Ush12,000 to 8,000 (about USD 6.5 to 5) following an increase in the number of charcoal traders especially the wholesalers using lorries. Thus economic incentives to conserve wood and charcoal have diminished over the past five years due to increased demand. The taxation on wood fuel serves no regulatory objective. That means that price of financial incentives to purchase more energy efficient stoves diminished.

5.7.2 Overall National Energy Policy

The energy sector has historically placed emphasis on policies that address supply of commercial sources of energy, giving little attention to biomass, which is the major source of energy in the country.

The main goal of the new National Energy Policy is to meet the energy needs of the Ugandan population for social and economic development on an environmentally sustainable way while the second objective is to increase access to modern affordable and reliable energy services as a contribution to poverty eradication. On the demand side the main objective for households and community is the provision of basic services including water supply and sanitation, health, education, public lighting and communication in order to improve the social welfare of the rural population.

The specific objectives for this sector are to:

- Achieve a sustainable level of energy security for low-income households so as to reduce poverty at household level;
- Improve the efficiency in the use of biomass resources, recognizing that biomass will remain a dominant source of energy, especially in the rural areas, for the foreseeable future;
- Specifically target provision of energy to productive activities such as home-based industries in order to directly raise household incomes; and
- Sensitize women on energy source and technology choices in order to reduce the labour, burdens and poor health conditions associated with biomass energy use.

The National Energy Policy recognizes the role energy supply improvement in rural areas is likely to play thus the need to include biomass in the realms of national energy planning.

It also recognizes that wood fuel harvesting contributes to degradation of forests as wood reserves are depleted at a rapid rate in many regions, the impact on the environment, health of end-users and the burden of collecting firewood on women and children as a result of increased use of biomass energy. In some parts of the country wood fuels are now scarce. This could be addressed to some extent through demand side management, which includes the use of energy efficient devices and alternative sources.

There is insufficient data on demand and supply of biomass fuels in the country and lack of awareness about the potential for biomass energy technologies. The policy views reinforcement of database on biomass especially the demand factors as a major thrust for planning purposes.

The policy recognizes the inadequacies within government institutions to plan for and monitor the sub-sector, and conduct research and development. The energy policy plans to increase private sector participation through use of smart subsidies particularly for improvement in efficiency and technology acquisition. Emphasis would be placed on improving efficiency of biomass use along the production to end-use chain.

The energy policy proposes strategies to address demand side in various sectors namely household, institutional, industrial, commercial, transport and agricultural sectors. Energy efficiency is the major thrust in demand side management. In addition the policy proposes strategies to address supply sub-sectors namely the power sub-sector, the petroleum sub-sector and the biomass and other renewable energy sub-sectors. The major objective of the biomass and other renewable energy strategy is provision of focused support for the development, promotion and use of renewable energy resources for both small and large-scale applications.

5.7.3 Forestry Policy and the National Forest Plan

Uganda's forests and woodlands are viewed as the mainstay in the three pillars of sustainable development; industry, society and the environment. The forest policy recognizes that Uganda's forest resources provide energy, supplying 93% of national energy demand.

Gazetted forest reserves, which cover about 40% of total forest area in the country are not the major source of biomass energy for current and future demand. About 35 million cubic meters of firewood are consumed annually. This is way above the total annual allowable exploitation of 350,000 m3 for all reserved forests. Since these forests still have growing stock, it implies that the bulk of biomass used for energy is mainly obtained from areas outside forest reserves. However, it is not documented as to how much of this quantity is supplied by gazetted forests.

The forest policy views farm forestry as major strategy in ensuring adequate supply of biomass energy. The role of the forest sector would be the provision of adequate, clean seed and advisory services to extension officers based at sub-county level. The planned extension service under National Agricultural Advisory Services (NAADS) is an effective way of reaching subsistence farmers thorough out the country. The policy also plans to take advantage of governments commitment to promote and develop farm forestry through the Plan for Modernization of Agriculture.

The use of forestry waste as a source of energy is not addressed but is implied since there are no planned energy plantations and so forest reserves are expected to supply some fuelwood as outlined in the National Forest Authority Business Plan. The tops and branches from logging operations would be sold as wood fuel. This wood fuel would be obtained from processing cycles as stipulated in the management plans for specific forest reserves.

The forest policy mentions collaboration with stakeholders such as poor rural and urban population, those working in wood industries, consumers of forest products, servants of the sector and the wider national and international public but no clear strategy is highlighted. Further, the policy specifies collaboration with the agriculture, land use, water, wildlife, industry and energy sectors.

5.7.4 Availability of Woody Biomass

The data on biomass supply assumes that all the biomass is accessible to the population. In reality, a lot of the biomass held in forests is either inaccessible due to long distance or because of management restriction in protected areas such as forest reserves, game reserves and national parks. For purposes of simplicity, if areas under protection are assumed to be unavailable or inaccessible, then the available supply of biomass for use shall be reduced drastically to about 50% of the total biomass available.

Sources of wood used for firewood and other forest products can also be derived from these distribution patterns and land use types. For example the nearest source for fuelwood would be the subsistence farmland areas or the nearby bushland areas.

5.7.5 Agriculture

Biomass energy from crops is derived from agricultural residues arising from the growing and harvesting of both food crops and cash crops. The most common food crops grown in Ugandan are Plantains: both green and sweet bananas; Cereals: finger millet, maize, sorghum, rice and wheat; Root crops: sweet potatoes, Irish potatoes and cassava; Pulses: beans, field peas, soya beans and pigeon peas; Others: ground- nut (sometimes called peanuts), and simsim. The cash crops grown are coffee, cotton, tea and sugar cane. The distributions of these crops in the country depend on specific crop requirements such as soils and climatic conditions (e.g. rainfall patterns and its distribution). The fol- lowing section presents the areas and production of the major crops in Uganda based on the 1995 data from the Ministry of Agriculture.

5.7.6 Agricultural Residues (Food and cash crops)

Agricultural residues are the materials left after harvesting or process- ing of crops. Data on residues are in most cases lacking. Therefore, the data for residues were derived by calculating the production figures of each crop with appropriate conversion factors (residue-production ra- tions). The total agricultural residue is about 8,000,000 tons of which about 1,734,000 tons is available for energy use.

5.7.7 Livestock

There has been no recent detailed livestock census carried out in the country. Thus the figures given are derived from the Ministry of Agri- culture, Animal Industries and Fisheries statistical abstract 2000. It was estimated that there were about 5,000 cattle, 900 tons of dry mat- ter per year of which about 240 tons is available for energy use.

5.7.8 Total Biomass Production and Use

The overall potential stock of woody biomass in Uganda in 1995 was about 477.2 million tons air dry. However if the biomass from protected areas were removed from the gross national stock, the net available stock would be reduced to 275.9 million tons air dry.

It should also be noted that in reality only small stems, twigs and branches are normally utilized. If this is taken into account, the avail- able stock would be further reduced to about 30% (about 90 million tons of wood).

The total sustainable biomass production and consumption flows have been summarized at national level but the information is so sketchy that it cannot be used for effective planning. The concept of sustainability is that the annual consumption should not exceed the mean annual increment. The annual gross sustainable yield at national level is about 20.4 million tons per year. However if the available supply from non-protected areas is considered alone, then the yield will be reduced to about 14.5 million tons per year.

As expected from the above presentations, the highest biomass supply was from trees (14.4 million tons per year, air-dry biomass above ground), followed by agricultural residues (1.7 million tons per year) and the rest were from animal wastes.

On the consumption side, firewood consumed by household constitutes the greatest amount of biomass, followed by charcoal, firewood for commercial purpose and residues. Only 50% of crop residues available for energy purpose are consumed. An increased supply from this source should be considered with care.

The data on land cover (use) distribution, areas, standing stock (biomass density), agricultural crop areas and production and animal waste production were used to quantify the supply of biomass. While data on charcoal and firewood from available literature were used to determine the quantity of biomass consumed in 1995.

On the supply side, Uganda had a total stock of 4772 million tons of woody biomass in 1995. The total biomass yields were about 20.4 million tons per year of wood (annual mean increment), 8.0 million tons per year of agricultural crop residues and 4.7 million tons per year of animal wastes.

The available (excluding protected areas) amounts for energy use were 275.9 million tons of wood as stock, 14.46 million tons of wood per year, 1.73 million tons of residues per year and 236,000 tons of animal wastes as yields.

On the consumption side in the same year, a total of 20.2 million tons was consumed of which nearly 80.5% was firewood, 14.5% charcoal and 4% residue.

Biomass flows based on these sustainable yields gave a negative balance of 3.8 million tons per year in 1995. The deficit for woody biomass is even higher: 4.9 million tons per year for the same year.

Some gaps exist in the data and the estimates of the above biomass should be treated with caution since many issues that affect the gross supply and the consumption were not factored in. Nevertheless, the general conclusion is that with a negative balance of 3.8 million tons, there are serious challenges to the sustainable use of this resource. In certain regions and districts the situation is even worse. Under the present circumstances, the future biomass supply for Uganda cannot be guaranteed. This is because the demand for biomass resources would intensify as the population increases. Remedial actions have to be put in place before the situation worsens.

5.7.9 Some Biomass Energy Challenges

This section reviews some biomass energy initiatives in the country and examines some of the barriers to dissemination of Biomass Energy Technologies (BET). The initiative mainly cover creating awareness and dissemination of technologies for tree planting and energy efficiency, particularly improved stoves and improved charcoal production techniques. There are some activities addressing capacity building as well. Most non-governmental organizations (NGO) and public sector initiative receive donor support while private sector activities are supported by the parent companies.

The Ministry of Energy and Mineral Development (MEMD) with support from development partners is implementing the Sustainable Energy Use in Households and Industry (SEUHI). The objective of the project was to improve efficiency in energy conversion and use in the households and small-scale industry. The project address rural and urban household stoves in Kampala, Soroti, Adjumani, Kabale and Tororo; charcoal production in Luwero, Nakasongola and Masindi; and lime production in Kasese, Kisoro and Tororo.

The MEMD has:

- Disseminated improved cookstoves in 34 sub-countries in the districts of Kabale, Tororo, Soroti and Adjumani. Over 400 people have been trained and about 7,000 households now use improved stoves.
- About 70 artisans were trained in production and marketing of improved charcoal stoves in Kampala and Kabale.
- Over 130 charcoal producers in nakasongola, Luwero and Masindi Districts were trained in improved charcoal production methods.
- Three charcoal producers associations were formed.
- Two charcoal producers associations benefited from a modest revolving fund through association's access to Kampala markets. The two associations have been operating stalls in Wandegeya and Namasuba Markets since August 2000.
- An improved limekiln in Tororo was completed and launched and another lime kiln was built in Kisoro.

- Energy audits have been carried out in some biomass-based industries.
- Eight tree nurseries were established in sub-counties in Kabale District.
- In collaboration with some agro-based institutions, MEMD planted over 500,000 seedlings in Adjumani District.
- Provided facilitators to Nyabyeya Forestry College on energy issues.
- Established 20 biogas digesters and trained 20 artisans in biomass digester construction.

Under the Energy Advisory Project supported by a European development partner, MEMD prepared the national energy Policy, sub-sector-strategies and energy Information System. Technical assistance was also provided to different NGO projects and the media. A Biomass Energy Development Programme (BEDP) was prepared. The MEMD plans to remove barriers to increase biomass energy efficiency in households and small-scale industries in rural and peri-urban areas. This will be done by promotion of improved technologies, energy saving methods and fuel substitution. High priority will be given to urban households, institutions and small industries where management and motivation to cut fuel costs would quickly settle in.

Biomass is used to supply energy for a number of activities. Some of these such as tobacco curing encourage the users to also plant trees as they consume available biomass materials. Several local NGOs and groups, such as Joint Energy and Environment Project (JEEP), Integrated Rural Development Initiatives (IRDI), Renewable Energy Development Centre (REDC) and learning institutions have collaborated with various international development organizations to promote tree planting and use of efficient energy devices.

5.7.10 Information Deficiency

There is lack of comprehensive and reliable data especially on the consumption side. Biomass energy planning relies on a wide variety of information (both quantitative and qualitative) from different disciplines. The responsible institutions such as Energy Department and National Bureau of Statistics are chronically handicapped by financial and human resources and usually occupy a marginal role in the national planning context. Even basic data such as the prices of woodfuel in different parts of the country may be missing or inconsistent. Quite often, the discrepancies among independent estimates illustrate the low consistencies and reliability of data sources, with obvious negative consequences for formulation or identification of priorities and definition of policies and strategies. The figures and facts about the central role played by the sub-sector in term of offering employment or its contribution to the country's GDP are not documented. If valid arguments are to be won in favour of the sub-sector, say for more resource allocation, facts and figures should obviously back such an argument.

5.7.11 Technology

The sub-sector is characterized by few producers/manufactures of proven Biomass Energy Technologies (BET). A large number of manufacturers produce items of low quality and sometimes worse than "traditional" technologies in terms of efficiency, durability and other attributes. There is lack of knowledge about the availability of new technologies, and the advantages and constraints associated with such technologies. Advanced biomass technologies such as gasification, and modern brick kilns are little known in the country. There is clearly lack of quality control and regulation in the sub-sector.

5.7.12 Institutions

The key institutions in the biomass sub-sector including MEMD, NGOs and private companies experience shortage of qualified personnel and expertise in the area of biomass. These organizations do focus on many other issues and biomass is more often a small component of their activities or programmes. This situation is aggravated by the fact that the institutional linkages between the various actors are weak and in some cases non-existent. There is need to enhance collaboration and exchange of experience between all the actors. The situation if further aggravated by the fact that financing mechanisms are not available for the BET. Many Micro Financial Institutions are not yet convinced of the potentialities of these technologies for the development of the rural populations.

5.7.13 Objectives of Biomass Energy Development Strategy

An ideal biomass energy strategy should ensure that resources are used without negative social, economic and environmental consequences.

This requires balancing the supply and demand of biomass energy. While the energy sector is responsible for energy supply and demand, a number of other sectors have a niche in biomass energy issues. These include forestry, agriculture, environment, industry, health, population, gender and education. Various other stakeholders are involved in supply and demand aspects of biomass.

The Energy Department has developed a National Energy Policy. The major strategy for implementing the energy policy in end-use sectors is energy efficiency. The biomass (and other renewable) supply sub-sector's objective is for "Government to provide focused support for development, promotion and use of renewable energy resource for both small and large scale applications". The policy provides a list of strategies, which need to be developed further in order to address the needs of the sector. The National Forest Plan emphasizes biomass energy conservation through "Developing a biomass energy strategy, improving uptake of energy efficient technologies and developing appropriate technologies for production, processing and energy consumption". The National Environment Action Plan also emphasizes energy efficiency, increased production of trees and increased use of alternative energy sources.

The major objective of this initiative is to synchronise and build on the strategies from different sectors that address biomass energy demand and ensure that stakeholders of the sub-sector play roles based on their comparative advantages in formulating and implementing the biomass energy demand strategy.

The issues to be addressed in this strategy are:

- Institutional Arrangements;
- Human Resource Development;
- Mobilizing Financial Resources;
- Dissemination and Awareness;
- Quality Control; and
- Research

5.7.14 Institutional Arrangements

Biomass energy should not be viewed as an isolated sub-sector but as an integral part of the development process. Uganda's economy is agro-based and biomass plays a key role in a number of agro-processing industries; therefore developments in the biomass energy sub-sector will contribute towards the Plan for Modernization of Agriculture (PMA). Biomass plays an important role in ensuring food security because it is and will continue to be the major source of energy for cooking for the next decades. In some woodfuel deficit areas, households have shifted from more nutritious foods that take long to cook to less nutritious easy-to-cook foods.

Apart from existing potential, biomass can play a role in power generation for rural based industry (rural electrification) thus contributing towards industrial development.

A revision of the pricing and taxation policies for woodfuels is a condition for the attainment of the policy targets. The level of taxation has to be raised step by step to promote management of the resource base. The primary objective of taxation policy for household energy should be to correct market imperfections that prevent prices on the market to reflect the correct economic cost of fuels. This would convey the correct price signals to the consumer of the cost of his fuel use to society and encourage a switch to the use of improved technologies and other energy sources.

Innovative institutional strategies are required in order to ensure that the energy needs of end use sectors that rely on biomass energy are met. Important institutional options include but not limited to:

- Careful review of the existing institutional framework to identify opportunities for rationalization and improve efficiency. This would include a functional analysis of agencies involved in the biomass energy sub-sector. In addition an analysis of the strengths and weaknesses of the existing institutional framework would help identify areas that need strengthening or improvement.
- Giving preference to simple policy instruments that yield substantial results at low cost.
- Formulating simple regulatory and fiscal measures that are commensurate with local enforcement and monitoring capacity. These measures should be based on realistic and technically proven strategies with stakeholder participation. For example licensing charcoal producers and fuelwood dealers at village, parish or sub-county levels would be best effected when local authorities have been involved in preparation of environment action plans which incorporate energy at local level.
- Using rescarch results to guide decision-making process.

- Incorporating energy planning at lower levels, mainly at district and sub-county levels.
- Undertaking regular and periodic reviews of past biomass energy policies, projects and initiatives to ensure that lessons learnt are incorporated in current and future activities.
- Ensuring that government departments focus on their central functions of regulations, evaluation and monitoring.
- Establishing appropriate pricing schemes for wood fuels to allow full recovery of cost for energy crops thus crating a favourable environment for investment in commercial energy plantations.
- Strengthening institutional database, proactive advocacy and reinforcing effective national coordination.
- Encouraging the establishment of energy service companies (ESCo) with special interest in promotion, testing, manufacture and marketing of energy efficient technologies and practices.
- Encouraging the development of professional energy-related associations.
- Encourage inclusion of biomass issues in rural development projects.

5.7.15 Human Resource Development

Successful energy programmes in Sub-Saharan Africa are those in which local initiatives played a dominant role in the project conceptualization through to implementation. Meaningful development of the biomass energy sub-sector will require mobilization, strengthening and effective organization of human resources in the country. Limited local participation has retarded the growth of local skills making Uganda more dependent on external expertise. In the short term, optimum use of existing skills would yield the highest benefits at the lowest cost. It is therefore necessary to establish the existing human resource capacity in the country. Over the years, there has been great dependence on external expertise in the energy sector. Appropriate long and short-term training should be made available to ensure that those employed in the biomass energy demand sub-sector can effectively participate in various aspects of biomass energy production, distribution and use. Among end-users, there is limited knowledge of biomass energy efficiency opportunities and availability of new technologies. Little is known about the capacity to undertake energy efficiency analysis and the simplicity with which energy savings can be used to increase production and profitability in small and medium enterprises.

Some tertiary institutions such as the Faculty of Forestry and Nature Conservation at Makerere University and Nyabyeya Forestry College have incorporated biomass energy in the forestry training curricula. This is the beginning of a process that will supply trained cadres for the sub-sector. In order to accelerate technology adoption rates, it is imperative to include biomass energy conservation in other tertiary institutions and at lower education levels. Some aspects of forestry and environment conservation are now part of the syllabus in primary and secondary schools in Uganda. It is envisaged that the younger generation will implement what they learnt in school. This could also be done for biomass energy conservation.

A number of institutions such as Nyabyeya Forestry College and Makerere University Faculty of Technology are planning to establish "Biomass Energy Centers" in different parts of the country. In addition 12 Agricultural Research and Development Centers (ARDC) (former District Farm Institutes) currently under National Agricultural Research Organization management and Agricultural Development Centers (ADC) under Local Government management can be used as technology uptake pathways for biomass energy technologies. These efforts need to be coordinated to avoid duplication.

A few formal and informal entrepreneurs have invested in production of biomass energy technologies especially production of improved cook stoves. They often have limited technical and business management skills. As a result, these businesses have not registered the success originally expected. Different small projects plan to provide training in both technical and business management skills for improved charcoal stove producers. Other small and medium scale entrepreneurs in the sub-sector need training as well.

In order to create a critical mass of personnel to implement the biomass energy strategy it is imperative to:

- Conduct a human resource survey among major actors and agencies to establish existing human resource capacity. This will help establish the gaps and thus plan for relevant training programmes. In addition, a training needs assessment should be conducted among stakeholder.
- Develop required human resource capacity to implement the National Biomass Energy Demand Strategy by:

- – Providing formal and on-job training for all levels of personnel in biomass energy projects. Training courses should be based on identified local needs of the target groups.
- – Training extension workers in relevant biomass energy technologies. Under the NAADS system extension staff hired by the district authorities will be the major link to farmers. Such training would be conducted if the farmers have identified biomass energy conservation technologies as necessary input to their agricultural development activities.
- – Training staff and managers of small and medium enterprises in technical and business management after training needs assessment.
- • Establish a network of trained biomass energy auditors who will be skilled in identifying and implementing profitable biomass energy efficiency initiative.
- • Integrate biomass energy conservation in the school curricula at primary and secondary level.
- • Co-ordinate biomass energy centre activities and establish demonstrations at ARDC and ADC.

5.7.16 Mobilizing Financial Resources

The level of dependence on external sources to finance energy development in Uganda is so prevalent that little thought is given to mobilizing local financial resources. Uganda has borrowed massively to finance the power sub-sector. On the other hand, the biomass sub-sector, which is an important source of livelihood for Ugandans mainly benefits from small grants. Mobilizing local financial resources is a pre-requisite for sustainable energy development in the country. A few private sector biomass energy initiatives have been implemented using locally generated resources however, experiences on the success of these initiatives is limited. Biomass energy supply is totally financed by local capital however it is characterized by under-valuation of the resource. Some financial institutions administer a number of special loan programmes, which are dedicated to the development of small and medium enterprises. Private sector agencies involved in biomass energy should take advantage of these loans.

Some biomass energy technologies such as biogas systems and institutional stoves; and, technologies for harnessing alternative fuels have high investment costs, which discourage end-users from acquiring these technologies. The first biogas digesters introduced in Uganda were made of concrete and/or bricks, which most farmers could not afford. Currently Integrated Rural Development Initiatives is disseminating a tubular biogas digester, which is cheaper. However, farmers still need financial assistance in order to purchase the biogas system. High investment cost may also affect the adoption of liquefied petroleum gas despite the fact that the industry has introduced cheap 5 kg cylinders with single burners.

Effective financial resources mobilization can be realized by:

- Developing mechanisms for mobilizing local capital to finance biomass energy projects. An Energy Development Fund could be established with contribution from a given percentage of tax revenue from all energy suppliers. Percentages could depend on volume of business and levels of pollution from the fuel.
- Encouraging greater involvement of local banks in biomass energy investments similar to the arrangement for acquisition of solar home systems. Small and medium enterprises should take advantage of loans provided by banking institutions.
- Include smart subsidies for biomass energy initiatives as part of the already established Rural Electrification Fund.
- Rationalizing the collection of revenues, fees and taxes related to wood production, transformation, transport and marketing at local levels. Those funds should be utilized to ensure maximum benefits are realized at local government level while promoting the commercial woody biomass sector to operate on a sustainable basis.
- Developing appropriate financing packages for small and medium enterprises in biomass energy projects.
- Promoting technology acquisition especially for alternative fuels through hire purchase system. This would be applicable for charcoal production technologies, institutional stoves, biogas and liquefied petroleum gas. In addition, revolving fund systems to enable farmers acquire tubular biogas digesters could be established. However, an assessment of these end-user financing systems is required before wider use.

5.7.17 Dissemination and Awareness

A number of agencies are actively involved in awareness and dissemination activities particularly for improved cookstoves and tree planting.

For most biomass energy consumers, reducing energy costs at end-use has not traditionally been a principal concern. There is need to make information more readily available to enable biomass energy users incorporate energy efficiency in their daily practices. While biomass energy technology dissemination efforts have been going on by different actors over the years, some of these agencies are bound to have registered successful dissemination. It is imperative to learn from experiences of different actors. It is also necessary to consolidate dissemination efforts and channel activities to where there is great need. The different media (press, radio, and television) and the professional associations concerned with Environmental and energy issues will have a crucial role to play in preparing and disseminating the information.

Innovative dissemination and awareness approaches will include:

- Designing a national awareness campaign to demystify BET and ensure actors are aware of key components of the biomass energy demand strategy. The awareness campaign would among other things provide information on biomass energy saving methods including different technology options, major actors and sources of information. The different media and the journalists associations will be trained to disseminate the right messages.
- Creating a database with information on technologies available and their performance parameter, institutions and their roles in the National Biomass Energy Demand Strategy, past, present and future projects and any other information considered necessary.
- Documenting and publishing reviews of technology dissemination activities so that other actors may learn from success and failures of past projects.
- Consolidating dissemination activities implemented by different actors to prevent duplication of efforts and concentration of activities in a given region of the country. This will help identify gaps in current dissemination approaches. Combined with data on biomass growth, demand and supply the dissemination approach would include identification of areas most suitable for implementation of biomass energy programmes.

5.7.18 Quality Control

Poor quality of some Biomass Energy Technologies is a barrier to large-scale dissemination. This is particularly prevalent in improved charcoal stoves but may also affect other biomass energy technologies and so there is a need to stop the trend before it spreads too far.

It is proposed that:

- A National Energy Reference Center be developed at Makerere University where some limited testing facilities for improved stoves have been established.
- A regular testing programme be instituted to ensure that acceptable qualities of goods and services are maintained.
- Quality guidelines and standards for production of different biomass energy technologies should be developed and technology producers provided with the information.
- Technology producers be encouraged to use trademarks on all their products.

5.7.19 Research

There has been limited support to research in biomass energy technologies. Research activities have mainly concentrated on charcoal production and improved cookstoves. Uganda mainly depends on research conducted in other countries. Continued reliance on technology development done in other countries is expected to persist for many more years but effort should be made to reduce it. It is however imperative that technologies are adapted to suit local needs. Research should therefore focus on:

- Collecting and processing relevant data for the implementation of the biomass energy strategy. Data on biomass availability is an important input to planning at lower levels.
- Incorporating environmental costing in analysis of different energy options to ensure that the realistic cost of biomass energy is applied.
- Developing national energy efficiency performance data, management regimes and relevant technical specifications for different biomass technologies.
- Developing or adapting technologies for biomass uses where improved technologies are not yet available in the country.
- Exploring the possibility of using biomass for electricity generation and also as a source of liquid fuel that can be used to blend imported oil-based fuels.
- Promoting energy policy analysis research to assess the impact of different energy mix strategies.

5.8 Concluding Remarks

The development of energy resources is to a large extent dependent on technological knowledge base required for innovativeness and scientific ability to adopt new energy technologies to suit local circumstances. Energy provision is by nature an expensive service and therefore the consumers must have the economic power to afford it. Chapter Five presents basic scientific and technological information required for the development of various energy resources and describes the processes of producing different energy conversion devices. The information, on one hand, provides the justification as to why these conversion devices are expensive and, on the other hand, allays the fear in developing countries that these processes are too complicated and therefore not easy to accomplish under the prevailing local circumstances. The chapter is the longest in the book and is intended to encourage researchers and policy makers to direct more attention and energy in the development of locally available energy resources. In this regard, a deliberate attempt has been made to limit theoretical knowledge in favour of more practical information. Theoretical knowledge is however provided where it is considered necessary in order to appreciate practical aspects. The following references have relevant materials for this chapter [2, 10, 13, 25, 26, 27, 38, 46, 47, 48, 50, 56, 58, 64, 65, 66].

6

The Status of Renewable Energy in East Africa

6.1 Introductory Remarks

We have identified energy sources in East Africa and noted that vital and economically crucial energies are fossil-based. Most of other forms of energy including electricity can be obtained by using oil-powered machines. Unfortunately East Africa is not endowed with oil and has to rely on imports that take a large fraction of her foreign earnings. Hydropower and geothermal sources can only produce electricity to light homes and provide power for lighting and operating machines for various applications. They cannot perform all the functions that oil does, particularly in the transport sector. Thus to fully meet energy requirements in East Africa, both oil and other sources of electricity will continue to be used. However, oil remains the most crucial source of energy due to its versatile sectoral applications. It can provide electricity, heat energy, light, and facilitate transportation for which it will continue to play a crucial role for many years to come. However, it is possible to restrict its use to sectors where there is no alternative. Energy for heat, lighting and home entertainment can be obtained from other local sources and in this regard, top priority should be given to sources that are readily available with high possibilities of sustenance. Energy resources that traditionally have been used in East Africa for many years should receive more attention as they have proved that it is possible to sustain their supply. These are biomass, solar, hydro and wind energies. The use of most of these sources has never been officially planned and therefore individuals and small non-governmental organizations with limited specific objectives have haphazardly handled their development and promotion. Thus, lack of coordination and institutional support has inhibited the development of technologies that would efficiently harness these energies. The new approach will have to deliberately plan and coordinate technological developments for these resources. Through past efforts of non-governmental organization and donor agencies, significant knowledge on options for disseminating Renewable Energies in East Africa has been accumulated but the results of many technological development efforts have been far below expectations. This has been so largely due to the fact that selection, development and implementation of appropriate technologies were not effectively supported by both legislative and institutional arrangements. It appears the governments did not recognize the complexity of this process and had hoped that, with the accumulation of knowledge and experience, adequate technology transfer would also take place. Unfortunately this did not happen and opportunity that would have pushed the region to a higher level of development was wasted.

In the 1970s when the oil crisis revived global interest in alternative energies, almost all countries that were severely affected started, at nearly the same level, to conduct research in the development of technologies that would harness solar, wind, and biomass energies more efficiently. If sub-Saharan Africa, in general, paid as much attention to these researches as did the developed countries, the situation would be much different today. It is indeed sad that countries like Kenya, Uganda and Tanzania which are located right on the equator where there is abundant sunshine throughout the year are now importing solar energy devices from other parts of the world that receive very little sunshine. It is equally sad that these are the countries whose 90% of the population still depends on rudimental use of fuel wood without any effort to develop or use modern technology for biomass energy conversion despite the growing demand for it in the region. These are the same countries with landscapes and highlands where there is abundant un-tapped wind energy. It is known that wind speeds increase with altitude and traditionally people in East Africa do not farm or live on the mountains and high hills. These are areas where wind generators would today be generating power to light the dark and poor villages in the rural areas. Because of the cool climate and fertile farmland around the hills and mountains, the population density is usually high and so many people would benefit from such wind generators. The problem is neither ignorance nor lack of resources. Many of the leaders and local technocrats are fully aware of the opportunities in Renewable Energies since they have interacted very closely (mainly through higher education) with countries like Germany, USA, Denmark, United Kingdom etc, which are the world leaders in renewable energy technologies especially wind energy. Furthermore, some locally produced wind machines have been operating successfully in East Africa, indicating that there is a reasonable level of awareness in the region.

With regard to availability of investment resources, these technologies are not as capital intensive as some of the developments that have been undertaken by the governments. Biogas digesters, Gasifiers, Wind Machines and Solar cell production plants are not as expensive to establish as, for example, paper, sugar or steel mills and large hydropower stations, which are found in the region. Renewable energies are the sources of energy on which the majority of the people have relied for centuries and yet they are the resources that the leaders have chosen to neglect. This is not to say that direct government involvement in this process is a prerequisite for their success. In some cases, government's direct participation may even be detrimental to the development of Renewable Energy Technologies but it is very important for the governments to provide visible and attractive incentives through appropriate legislations and institutional framework. In addition to this, the government, through her extensive service delivery network can be the largest consumer of these technologies by formulating and implementing promotional policy that requires all government institutions like schools, rural housing schemes, dispensaries and hospitals to use appropriate renewable energy sources. The importance of energy in any development process particularly in rural development should be recognized and given the emphasis and support it deserves by establishing institutions that would oversee renewable energy development and application. The existing power utilities in the region have not taken up this responsibility and, given their interests and limitations, it is doubtful whether they can be effective in promoting the development of renewable energy technologies.

In order to diversify energy production in East Africa, it is important to consider the end-use and then choose appropriate source for this. If the end use requires heat only then there is need to choose the source that would efficiently provide that heat and if the end use needs light load electricity then the source should provide just that and the appropriate ones in this case would be solar, small hydro, wind or producer gas-powered generator. It is also important to analyze the number and concentration of the consumers as this will dictate whether the source should be centralized or decentralized. Obviously, for the scattered distribution of rural settlements in East Africa, the most suitable sources for energy supply would be those that are scattered and decentralized just as the human settlements. This arrangement would not only solve the problem of high transmission cost but would also involve the communities in the ownership and management of the facility. Such community participation are some of the characteristics for sustainability that can only be provided by renewable energies such as small hydro, solar, wind and modern biomass technologies (biogas, co-generation, and gasification). One major problem so far has been lack of organized data on availability and sites of various renewable energy resources. This makes it necessary for an investor to carry out data collection besides other feasibility studies - a tedious process that most investors are usually reluctant to undertake. It is therefore necessary for the East African countries to undertake an extensive renewable energy resources assessment at both regional and national levels and document them as accurately as possible, giving details of the type, location, quantities and daily, monthly and annual variations of the available energy resources. Additional well-documented information on sites that are economically and technically viable as well as socially acceptable should also be available to enable interested investors to make informed decisions.

The issues of quality control and safety requirements would be managed through appropriate legislation and institutional framework specifically established for the development and application of renewable energies. Previous attempts to promote renewable energy technologies in the regions severely suffered due to poor qualities and wrong unchecked maintenance procedures that seriously undermined consumer confidence. As a result most of them could not stand the test of time. This is an aspect that is threatening the survival of the Kenya Ceramic Jiko which is considered as a very successful renewable energy device in East Africa. Many manufacturers, taking advantage of the high demand and working without quality control enforcement, have used sub-standard materials to produce low quality stoves and this is rapidly eroding the consumer confidence. Installation of faulty or incomplete solar energy systems like solar water heaters and photovoltaic panels and the emergence of unscrupulous dealers who knew very little about the technical operation of the systems and consequently gave wrong information to the clients caused a drastic drop in demand for these devices in Kenya. Biogas plants constructed in the 1980s with a lot of support from donors are no longer in good working conditions because of poor management (irregular feeding, poor and delayed repair works etc). Similarly the fuel briquette machines and biogas plants which were promoted in Uganda in the early 1980s are either operating much below the capacity or are not working due to management and maintenance problems. There are many examples in which good renewable energy systems did not succeed because of inadequate technical support and distribution points as well as the absence of quality monitoring and enforcement mechanisms. Obviously rural conditions are also not favorable to traders and some problems arise due to too few dealers who find it too expensive to follow up scattered installations over large distances and consequently such installations once broken down remain so for ever. Installed wind machines for water pumping and electricity generation have been most affected. For example, in Kenya, the Kijito wind pumps are scattered and technicians have to travel long distances to maintain them and so the maintenance cost incurred by the user is inhibitive and discourages widespread use of the technology. The desire of the manufacturers of Pwani wind pumps to remain in the market forced them to sell their products within manageable distances so that they can serve their customers effectively.

At the moment, renewable energy devices particularly wind, solar and some biomass devices are thinly spread out in the expansive East African territory and the number of qualified maintenance staff are too few to carry out cost-effective repairs. The situation becomes even worse when the system are wholly imported so that the user does not have anybody to turn to for assistance. All these factors put a lot of constraints on renewable energy systems and therefore deliberate efforts must be made to effectively address them. One way to do this is to encourage local production centers for the systems that require initial large investments like wind machines and solar systems as well as the development of local manpower for maintenance and repairs. Manpower development should cut across all sectors of the society from artisans to university graduates with the capacity to continue with research and technical modifications. This will call for the development of suitable training syllabuses at all levels of education and training institutions. The need to set up institutions specifically charged with the responsibility of developing from within, and acquiring from without, renewable energy technologies that are appropriate to the different circumstances of the users is long overdue. It is only through such institutions that some progress can be made and evaluated.

The various ministries in charge of energy matters should have, as their main extension responsibility, renewable energy development section that should promote and coordinate research in energy technologies. So far there are a number of institutions responsible for various issues such as monetary studies, cooperative development, meteorology, mass communication and so on, but there is none for energy studies and development. Yet without energy no progress would be made in any of those other areas. Admittedly, there are already a number of renewable energy activities going on in research institutions and universities in East Africa, but these are very haphazardly done with almost no participation at the national level. In Kenya, Kenyatta University established the Appropriate Technology Centre to train students up to postgraduate level on various aspects of technologies with renewable energy technologies as the main focus of the programmes. The departments of mechanical and electrical engineering in other public universities are also engaged in research covering renewable energy technologies. University of Nairobi, University of Dar-es-Salaam and Makerere University are all doing research in various aspects of renewable energy technologies including solar energy materials for PV cell production. However, at the national or regional level, nobody cares what happens to the results of such academic research efforts.

It is very necessary to recognize the importance of energy when formulating investment incentives so that both the government and the private sector can work together to not only support but also give research contracts on renewable energies. It should not be left to the individual researcher to decide how and what aspect of energy to study. In spite of national governments' apathy, some local energy specialists have carried out useful studies that can form the basis of future renewable energy development strategies but, again, such information is of no use if there is no planned and focused programme for relevant further development. The first priority therefore is to establish a technical institution in charge of research on renewable energy resources and the development of appropriate technologies for their applications.

At present, national governments are supporting various research organizations that are managed as separate entities with very little, if any, inter-institutional collaboration. This is not the best way to manage research activities in a developing country that is operating under very severe resource constraints. Individual institutions are more expensive to maintain than when they are managed under technical coordination of one national body. In addition to this, isolated institutions operate under restrictive mandates that limit their flexibility to accommodate new research and development areas that frequently emerge in developing countries and, as a result, new institutions are established every time there is need to address an issue which is not covered by the mandates of existing institutions. The developing countries have and will continue to meet new developments issues to deal with and therefore there is need to reorganize research and development institutions in such a way that they can cover these new areas as they emerge. They can however be clustered into groups of related fields in order to improve internal interaction and sharing of expertise and facilities. For example, the broad groups could be: socio-economic development, Science and technology development, governance and policy development, etc. All scientific and technological concerns would then be covered by, say, National Institute for Science and Technology, managed by high caliber scientists and covering research and technology development in issues such as energy, medicine, agriculture, veterinary, materials, etc. Such broad-based centers would be used and, if necessary, contracted to develop specific technologies that would enhance national development. The management of such establishments would have collaborative mechanisms with universities in which there is a vigorous exchange of expertise under sabbatical or secondment arrangements. This is the only way to efficiently and effectively use limited human resources and research facilities that are required at that high level of operation. This arrangement should ensure that energy research and development finds its right place in the region. Once these suitable institutional frameworks are established, the energy concerns should focus on local energy resources and use the existing potentials in ethanol production, co-generation, small hydropower, solar, wind and biomass to develop technologies that would lead to use of cleaner energies in the rural areas. The primary aim should therefore be to gradually reduce the use of raw biomass as the main source of energy in the region by employing a deliberate and well-coordinated effort to gradually and systematically introduce its replacements. In order to succeed in doing this, some attention would have to be directed towards changing energy consumption pattern in the region.

Ethanol production potential should nationally or regionally be developed to address the problems that the region is likely to face due to the imminent oil scarcity in the future. Developments of solar, wind and small hydropower should principally be geared towards improving rural electrification. In this regard, solar energy applications can be approached from two angles: use of photovoltaic technology to produce electricity, and, use of solar concentrators to heat suitable fluids that can produce steam to run turbines for electricity generation. Technologies for both these approaches are still advancing but the currently known processes can be adapted to the local conditions and used to produce electricity at reasonable efficiency levels. Thus it is worth considering some investment on modest solar cell manufacturing plant in East Africa given the long-term benefits that the region will have and the uncertainty surrounding future supply of other fuels. In the early 1990s, an analysis was carried out that compared the cost of solar electricity with that of grid power and it was established that the lifetime cost of solar electricity was about 25% lower that grid electricity. Since then, the price per unit peak power of solar photovoltaic panels has gone down, making power from this source more attractive. Application of other sources of energy should also be diversified, for example, biomass technologies could also concentrate on producer gas, biogas, fuel briquettes, co-generation, ethanol production and organized and efficient methods of charcoal production.

6.2 Renewable Energy Research and Development

While technologies that harness fossil and hydro energies are, on average, well developed world-wide, renewable energy technologies are not well developed in East Africa. Moreover, the distribution of electricity and oil are such that the rural populations do not have instant or easy access to them. In addition, the purchase and maintenance costs of the technologies that utilize these resources are too expensive for the people. Conversely, light and heat energies from the relatively readily available biomass resources using cheap and low-level technologies can easily be harnessed.

Practically all East African countries spend less than 1% of their GDP on scientific research. Consequently, very little resources, if any, are spent on renewable energy research and development. Most activities related to renewable energy technologies are externally funded and are usually initiated by fresh researchers returning home from overseas training programmes. In many cases, the projects are merely a continuation of research work done abroad and may not have any significant impact on the country's renewable energy development programme. Gradually these scientists are "de-activated" by widespread internal bureaucracy and inadequate financial support, which force them to turn to other activities. In spite of this, a modest number of researchers in the universities are conducting serious research on renewable energy technologies. The private sector is mainly engaged in commercial production and installation of technologies such as solar water heaters, biogas plants, wind turbines, charcoal and wood stoves. Governments or parastatal corporations control electricity generation by geothermal, hydro and large oil-fueled generators. Other agencies for example, NGOs are involved in extension work and rarely in research and development.

6.2.1 Biomass Energy

Biomass is the main, and indeed, traditional source of energy for the rural domestic sector in practically all countries in East Africa. The first growing population of the region is constantly increasing the demand for wood fuel so that forests are consumed much faster than they are regenerated. It is therefore necessary for the people to manage the wood-fuel resources properly for it to remain renewable. During the last two decades, deforestation for wood fuel has been a major concern for most non-governmental organizations. Subsequently, they have undertaken several activities to redress the situation. These include: improved charcoal stoves, wood and sawdust domestic stoves; institutional stoves; energy conversion technologies such as wood-to-charcoal; rice husks to charcoal briquettes; animal dung to biogass and solid biomass to gas. Some of these activities have been reasonably successful. For example, the Kenya Ceramic Jiko (KCJ), a technology adapted from Thailand, which was first introduced in Kenya has successfully spread to Tanzania, Uganda, Ethopia, and Sudan.

One aspect of biomass energy that has not received sufficient attention, is the extraction of fuel-oil from energy crops. Research in this area will be important especially if the research for direct substitutes for fossils oil is to continue.

Biogas, also classified as a product of biomass materials, has also been researched into and developed. Although biogas plants are advantageous in that the by product is useful as soil fertilizer, the technology has not been successfully implemented on a wide scale. The initial investment cost is high and it requires qualified manpower and constant attention in order to operate efficiently. Tanzania has however demonstrated that some of these constraints can be overcome. But further work is required to explore the possibility of commercializing the gas instead of the plants.

6.2.2 Solar Energy

The technology of converting solar energy into heat and electrical energies is rapidly picking up. Already there exists a large number of solar water heaters installed in residential premises, hostels, hospitals and hotels. The application of the photovoltaic cells is also increasing but at a lower rate due to the high initial costs especially since they are mostly imported into the region.

For costs to fall, the challenge is two fold. At the development and production level, there is need to increase the energy conversion efficiencies and at the consumer level, the systems should be optimized. Unfortunately, research and production of solar cells require in-depth knowledge of structures of materials and fairly sophisticated technology. Therefore, researches that are done in the local universities are severely constrained by inability to acquire the needed high precision equipment.

6.2.3 Wind Energy

Through suitably designed mechanical linkages, wind power can be harnessed to drive water pumps or to generate electricity. However, wind is a highly variable source of energy because its speed changes with changing geographical locations, land features, altitude, time of day and season. Hence, the challenge in the development of wind energy technology is to design machines that can produce optimum energy safely and reliably. Although the basic technology is simple, meeting this challenge requires highly sophisticated technology built on aerodynamic principles and using advanced materials and electronics. Thus, to develop a good machine for a given set of conditions requires a lot of financial and technical support. Unlike solar devices, wind machines must be matched with the local conditions where the machines are to be located. These constraints have retarded the widespread development and dissemination of wind energy technologies in East Africa as little research is carried out in the region to address them.

6.2.4 Research and Development

At present, most of the activities in renewable energy technologies are of the extension type in which non-governmental and the private sector play the major role. Most of the funds for research and extension work come from external sources. Some research and development projects are conducted at the universities and national research institutions with hardly any collaboration with either the policy makers or funding and extension agencies.

The region's research and development programmes in renewable energy technologies are beset with numerous problems including insufficient manpower, lack of clear objectives, lack of analytical skills and facilities, and poor incentives for researchers. Besides, large corporations involved in energy related activities do not locally conduct or contract research work on the development of renewable energy technologies. This policy has alienated local researchers and stifled the development of local capacity for renewable energies. Most of these problems arise from weak and ambiguous renewable energy policies that most of the countries in East Africa have adopted. In fact, some of the countries operate on the basis of haphazard public political pronouncements.

On average, the region is amply endowed with both biomass and solar resources. Both of these can be suitably used as small scale decentralized sources of energy in the rural areas where, for cost and availability reasons, renewable energy systems may be the only option. However, the exploitation of these resources will depend, to a large extent, on the individual country's ability to create or strengthen institutions for this purpose.

Therefore, the lack of large scale adoption of these technologies should not be viewed as an indication of inappropriate conversion device but as a reflection of the inadequacy of their research and development. Moreover, for the full benefits of these technologies to be realized, it is important that researchers appreciate the user needs and match them closely to the existing technologies. Unfortunately, many research programmes in the region do not address these issues at the initial development stages.

6.3 Impact of Energy Use on The Environment

The earth is a large energy store and its ability to retain that energy depends on the composition of the atmosphere, which to a large extent, is influenced by factors associated with energy production and consumption. Changes in the concentration of atmospheric substances can create climatic conditions that would reduce the earth's ability to support life. Energy production and consumption produce waste products that are capable of up-setting the desirable concentration of atmospheric substances. Every energy initiative must therefore consider strategies that would protect the atmosphere so that it maintains suitable climatic conditions. In this regard, it is important to assess the level of emission of atmospheric pollutants due to energy use in East Africa. Examples are taken from Kenya since it is the largest energy consumer in the region. This information would assist energy planners to choose appropriate energy development pattern for the region.

For the atmosphere to be dynamic, as indeed it is, to maintain its composition, and also to produce suitable world climates, requires energy. This energy comes from the sun; all the necessary conversions and transfers are accomplished through the interaction of solar radiation and matter. These processes naturally maintain the atmospheric composition at a necessary level for suitable climate.

If the world did not have atmospheric gases, the average temperature of the surface would be about 253K (-200°C) but due to the presence of the atmospheric gases, the average observed surface temperature is about 288K (150°C). The earth is thus kept warm at a fairly steady average temperature by the greenhouse effect of the atmospheric gases whose composition must not be upset. But our reliance on various types of fuels is continuously increasing and this is affecting the natural balance of the atmospheric composition. In modern world, the production and use of energy determines the rate and level of development. Energy is the fundamental unit for the physical development of the world. It provides vital services for human life, for example, power for transport, mechanical work, cooking, manufacturing etc. This energy comes from oil, gas, coal, wood, nuclear and primary sources such as solar, wind, and water. All these must be converted to the suitable forms for various kinds of machines or any other type of equipment or application. However, every source has its own economic and environmental costs, benefits and risks. Choices must therefore be made with full knowledge of environmental and economic implications. The current trends of both constantly increasing energy consumption and the environmental risks of energy by-products are disturbing.

One of these risks is the observed indication that there is climate change due to greenhouse effect of gases emitted into the atmosphere. Carbon dioxide is the most significant contributor, alone accounting for about 50% of the combined effect of all greenhouse gases at the present time. The alarming fact is that currently there is no effective technology for removal of carbon dioxide emissions from combustion of fossils fuels. Recent estimates indicate that the accumulation of carbon dioxide and other trace gases in the atmosphere would lead to an increase of the mean global surface temperature of between 1.5 and 4.5 degrees centigrade by the year 2030 if nothing is done to stop the trend. The effect of this global warming would be disastrous.

Let us now confine the discussion to the emission of CO_2 into the atmosphere. As has been mentioned, carbon dioxide is one of the most effective greenhouse gases in the atmosphere and its increased concentration will adversely affect the climate. It is generally accepted that carbon dioxide concentration has constantly increased at the rate of about 4% per day during most part of this century. It is estimated that at this rate, about 5×10^{10}kg of carbon per year is added to the atmosphere. Deforestation and changes in land use are other sources of the increasing atmospheric carbon dioxide. Although oceans and forests act as partial sinks to the emitted CO_2, it is clear that continued use of fossil fuels will lead to an increase in the atmospheric CO_2 concentration. Climate observations and studies with climate models suggest that the doubling of CO_2 concentration would give an increase in the global average earth surface temperature of 1.5 to 3 degrees centigrade, all other factors being constant.

6.3.1 Energy Contribution to CO_2 Emission in Kenya

Most of the reports and predictions of a future climate change due to increased atmospheric greenhouse gases have been based on a combination of guesses and estimates -'guesstimates'- derived from available information at the time. It should however be appreciated that although these' guesstimates' are reasonable indications, their information base is rapidly changing as more data are obtained and our understanding of the atmosphere-ocean-biota interactions improves.

In the following paragraphs, predictions on the Kenyan carbon dioxide contribution will be made on the basis of recent estimates, which are available in the literature.

Globally, the developing countries contribute about 15% of carbon dioxide emitted to the atmosphere. Thus, from a global perspective, CO_2 emission from a single developing country like Kenya would seem to be insignificant. We must however remember that atmospheric pollution protection and hence protection of the climate is the responsibility of all countries in the world and it is this combined concerted effort that will save the world. Like in most parts of the world in recent years, the seasons in Kenya seem to be shifting unpredictably and the rains becoming less reliable. This may be the result of a global climate change due to increased greenhouse gases in the atmosphere! Malaria-carrying mosquitoes, which were once confined to the hot lowlands around lake Victoria and the coastal strip have appeared in the high-highlands of Kenya, which were known to be too cold for the insects. Clearly these highlands are either warming up or the mosquitoes have adapted to the cold climate.

It is known that the world is now committed to further warming due to increased greenhouse gases, which are already in the atmosphere. Carbon dioxide is one of the most important of these greenhouse gases, and in Kenya, its atmospheric concentration is enhanced, mainly by direct burning of fossil fuels, deforestation and wood fuel consumption. Kenya depends very heavily on these sources of energy. Wood energy is the most important source of energy, accounting for nearly 72% of all energy use with a large proportion of the population depending on woodfuel and charcoal. Most wood in Kenya is used for household cooking and space heating. This traditional application of woodfuel is a major source of CO_2 emission which is further enhanced by the high rate of wood fuel consumption which exceeds replenishment. This is compounded by the fact that Kenya has a very small area of forest cover compared to the total size of the country. The total area of Kenya is 560,186 km^2 out of which 11,239 km^2 is water surface (rivers, lakes and part of the Indian ocean coast line of about 400 km). Only about 15% of the total land area is used for intensive agriculture and supports about 65% of the rural population . It is in these small areas of Kenya where most forests are found.

High population growth rate coupled with high fuel wood demand and various changes in land use are causing deforestation, which leads to a reduction in the size of the already small forest cover. Deforestation increases CO_2 emission in three ways: first, forests act as sinks for CO_2 much of which is stored in the soil; second, the process of decomposition of forest materials produces CO_2; and third, the burning of wood is accompanied by large CO_2 emissions.

In 1988 an estimated total of 101,000 m^3 of fuel wood and charcoal was consumed in Kenya (CBS, 1989) while 780,000 m^3 of timber was used for other purposes such as construction and paper production. These figures were obtained from the commercial sector. However, the vast majority of wood is used as fuel wood and construction material through non-commercial channels.

Actual fuel wood consumption in Kenya is therefore estimated at about 0.6 tonnes per year per person. For a population estimated to be about 31 million in 2005, the total annual fuel wood consumption is about 18.6 million tonnes (one cubic meter of fuel wood is approximately 0.7 tonne), while to meet the demand Kenya would require over 30 million tones.

Another major source of CO_2 emission is from the burning of fossil fuels. In 1988 Kenya imported about 2,022,000 tonnes of crude oil which was refined in the country. Out of this, domestic petroleum fuels consumed was about 1,730,400 tonnes. Transport sector (road, marine, rail and aviation) is the main consumer of petroleum fuels taking about 1,165,000 tonnes in the same year and thus accounting for about 67% of the domestic petroleum fuels. Industrial application accounted for about 25%. Within the transport sector, road transport is the principal consumer of petroleum fuels and this is one area that needs to be seriously considered in order to reduce CO_2 emission into the atmosphere.

It is interesting to note that while crude oil importation remained fairly constant between 1984 and 1988, the proportion used in transport steadily increased from about 1,029,000 tonnes in 1984 to 1,165,000 tonnes in 1988. This was due to an increase in the annual number of newly registered vehicles from 15,694 in 1984 to 18,764 in 1988. It appears that the energy conservation achievements scored by industry and other government machinery were consumed by increased activities in the transport sector. Currently these figures are much higher as can be derived from data given in this book.

Despite availability of information on the use of fuel wood and fossil fuels, it is not possible to make an accurate assessment of the amount of carbon dioxide emitted by these sources per year and hence the magnitude of its impact on the environment can only be estimated. One thing is nevertheless clear: the rate of carbon dioxide emission into the atmosphere must be restricted to manageable and environmentally safe levels. This is particularly important in Kenya where the forest cover is estimated to be only 5% of the total land area.

Carbon production from fossils fuels as per 1988 oil consumption of 1,730,400 tonnes using conversion factors recommended in the literature is 1.5×10^6 tonnes, which is equivalent to 0.0013 ppm increase in carbon dioxide in the atmosphere.

Cement production is another source of carbon dioxide. The annual production of cement from the two plants that existed in Kenya in 1988 is 1.5×10^6 tonnes. This gives an estimated emission of 2×10^5 carbon with carbon dioxide concentration of 0.00005 ppm. A third cement plant, which has been commissioned has raised cement production level to more than double the 1988 figures. Thus the production of carbon from this activity is estimated to have increased two-folds by 2005.

Terrestrial emission from deforestation and other types of vegetation and wetland drainage cannot be predicted from available data as they depend on the type of vegetation, rate of destruction/decomposition and wetland drainage. Forests and other types of vegetation also function naturally as net carbon sinks because the production rate of organic matter exceeds the rate of decomposition. Surface water masses also absorb and store CO_2. The present uncertainties with estimating carbon dioxide absorption/emission rates by the biosphere, and lack of adequate information on the Kenyan situation make it difficult to quantify the CO_2 cycle in the biosphere. Consequently, in giving the data for carbon dioxide emissions in Kenya, it is assumed that all the un-accounted for CO_2 sources, such as decomposition of biomass materials, marine productivity, charcoal production and changes in land use, and carbon dioxide sinks such as surface water masses, forests and grasslands, balance each other and are therefore not considered in this analysis. These projections are therefore associated with uncertainties. Thus for the purposes of this analysis, the atmospheric carbon dioxide increase attributed to Kenya is linked to the combustion of oil and wood fuels (including charcoal) and the production of cement which produces a combined annual total of 6.87×10^6 tonnes of carbon. This is equivalent to 0.0018 ppm annual increase in CO_2 in the atmosphere. In global perspective, this is only about 0.1% of the total annual global atmospheric CO_2 increase of 1.5 ppm - too small to warrant a national effort unless it forms part of a wide international or worldwide undertaking. Thus the calculated annual contribution of Kenya to the global increase in the concentration of carbon dioxide is small compared to those of a developed region and the World. The Kenyan estimates as well as the others based on contributions from fossil and wood fuels and cement disregard unaccounted for CO_2 sources and sinks. Carbon dioxide emissions from the three major sources discussed above are very closely associated with population growth and therefore the situation is expected to change in future.

In order to predict the future situation, it is necessary to define the atmospheric increase of CO_2 in terms of the per capita Kenyan contribution, which we estimate for 1989 to be 0.286 tonnes. This is 25% of the global average per person carbon dioxide emission of about 1.16 tonnes. The population of Kenya would double by 2020 if the annual population growth rate remains at the present level of about 3%.

Consequently, with the present trends of CO_2 emission remaining unchanged, the total CO_2 production in Kenya will also double. This projection still gives an insignificantly small quantity of CO_2 emission in Kenya. It however, does not consider industrial expansion and deforestation as a result of new human settlements, construction and grassland conversion into agricultural land. But even if it did, the figure would still be small. The important fact to note here is that although CO_2 emission from some developing countries is negligible, the net carbon dioxide emission into the atmosphere will continue to increase and may increase beyond the predicted level if nothing is done to change the trend. Industrialized countries have in the past emitted large quantities of carbon dioxide into the atmosphere and continue to do so in spite of the many international CO_2 control agreements. For example, the state of Michigan with less than half of the population of Kenya had a total annual CO_2 emission rate of 4.26×10^7 tonnes of carbon in 1987. This was about 6.2 times more than the total CO_2 emission in Kenya. On the other hand, total CO_2 production due to wood fuel consumption in Kenya was 1.7 times higher than that of Michigan (8,540,500 tonnes), a fact which underscores the significance of wood fuel as a major CO_2 source in Kenya as in other developing countries.

Globally, the atmospheric carbon dioxide concentration has already increased during the last 100 years due to human activities, especially industrial development, burning/combustion of forests and other biomass materials, and energy consumption. This has caused the earth to warm up by an average of 0.5°C and, in fact, the earth is already committed to further warming due to the already accumulated greenhouse gases in the atmosphere.

It should be recognized that the atmospheric concentration of these gases will, in all likelihood, continue to increase resulting into a very destructive effect on the climate and global ecosystem. In recent years, many parts of the world, including Kenya, have more frequently experienced 'natural' occurrences including ecological changes and climatic disorder, which probably could be initial consequences of the global warming.

There is only one global atmosphere and it belongs to all nations of the world. Atmospheric pollution in California or Scotland will affect Kenya or any other part of the world. Every nation must therefore join in the worldwide effort to protect the atmosphere in order to maintain a habitable world.

Tropical rainforests, which are mostly found in the developing countries, seem to have the most intense rates of carbon recycling and their conservation upgrading of productivity will go along way in the management of the global carbon dioxide problem. It is clear that biomass is an important source of energy in Kenya as has been discussed elsewhere in this book and this is typical of most developing countries. For example, in 1988/89, the world's energy consumption was estimated to be about 3×10^{20} joules out of which oil contributed 39% while biomass (90% wood) contribution was about 14%. In developed countries 45% of the energy was supplied by oil, 26% by hydropower, 3% by nuclear and only 1% by biomass. But in developing countries, 44% of energy came from biomass and 24% from oil. These contributions vary from country to country but globally they imply that about two billion people totally depend on biomass energy. It is therefore possible to balance their contributions to the global carbon cycle by ensuring that production equals or exceeds consumption. To do so will require a concerted effort on the part of all peoples to adopt energy policies consistent with sustainability of supply through the continued development of renewable resource technologies.

6.3.2 Measures and Options

Kenya, along with other countries in East Africa, has realized the need to control CO_2 emission and is making an effort to reduce the problem.

Kenya recognizes the important role of forests as a CO_2 sink and is emphasizing afforestation measures as a priority objective for facilitating sound environmental conservation. The Rural Afforestation and Extension Services has continued to implement afforestation programmes with the aim of sustaining a balanced ecosystem while making available forest products to meet the varied demands. In addition, agroforestry has also been encouraged.

On the control of CO_2 emission from the burning of fossil fuels, the government is encouraging conservation activities. However, at the moment there are no strict measures to ensure that afforestation and conservation activities are undertaken. The National Environment Secretariat (NES), which was formed to coordinate environmental protection activities has been carrying out environmental education and information activities, village assessments, and a national plan of action to combat desertification and to control pollution. Similar programmes exist in Uganda and Tanzania.

To reduce the emission of carbon dioxide from energy sources, the governments should recognize the risk to health and the dangers of global warming due to energy derived from fossil fuels and fuel wood. Strict energy efficiency and conservation measures must be applied in order to reduce atmospheric pollution by carbon dioxide. Since transportation activities in East Africa are the single largest source of CO_2, a reduction of emissions can be made through a combination of more stringent emission limitations for cars, buses, and trucks, strengthened inspection and maintenance procedures, and the use of clean fuels in commercial and private vehicles. Measures that would reduce unnecessary use of vehicle and improve traffic flow would help reduce CO_2 emissions. Some of these measures include greater use of public transport, high charges for parking space and good plan for non-motorized vehicles (bicycles) on all new roads. Efforts to implement these measures in the future should be seriously considered. In addition to these measures, provision of energy must be carefully planned and effectively implemented. In this regard, renewable energy should be developed to reduce large scale use of fossil fuels. Many developing countries including those in East Africa are located within the tropics where there is abundant solar energy throughout the year. Solar energy technologies for the supply of electricity and heat energy would therefore be a very suitable option for these countries. So far the use of solar water heaters is rapidly gaining popularity in East Africa. Photovoltaic panels (solar cells) are also in the market but customer response is still very poor. Perhaps the government should introduce legislation regarding the use of solar energy in all new establishments (building, industries etc).Wind energy is another renewable energy, which has some potential in the region. There are a few local firms, which are already manufacturing and marketing wind energy devises. However, the use of renewable energies, in general, is still less than 0.1% of the total national energy consumption.

Non-governmental organizations have been supporting governments' efforts by promoting energy conservation and the development of renewable energy technologies. This support should continue and, where possible, should encourage the highest standards of pollution control in industries and in the transport sector, and promote the adoption of life styles and technologies, which are consistent with sustainable development for all nations. East Africa in its desire to use more energy for development should do so while taking precautions to avoid mistakes made by already industrialized countries in order to limit the amount of green house gases emitted into the atmosphere.

6.4 Concluding Remarks

The development, use and consumption of energy produce various types of wastes, which are normally discharged back into the environment. These wastes, if not properly managed or controlled, have very undesirable impact on the environment. The chapter discusses the status of renewable energy in East Africa, general impact of energy on the environment and possible measures and options available. Related information that supports the presented position is available in [3, 8, 17, 18, 19, 20, 22, 24, 28, 29, 30, 33, 40, 42, 45, 49, 51, 57, 60, 62, 63].

7

Options and Challenges

There are many challenges in making the right choices that would lead to higher level of self-sufficiency in energy. These include technical, institutional, legislative and socio-cultural considerations and for any approach to succeed, locally available resources must be the foundation on which the plans should be based. Experience and technical details can be sourced from elsewhere but the core resources must remain local. In this regard renewable energies particularly solar, biomass, wind and small hydro potentials, which are fairly well distributed in the region would play a very crucial role as energy resources not only for East Africa but also for those countries that are not endowed with conventional sources of energy. As has been noted, rural settlements in the developing countries are hardly planned and are therefore randomly distributed without any consideration to the fact that there may be need to share some facilities. Every family is on its own and struggles to make the best use of whatever resource is available within easy reach. This style of living does not encourage the development of sophisticated and efficient techniques that require the concerted effort of members of the community. As a result, people tend to continue with the old traditional and often inefficient methods of meeting their needs and therefore the efforts are usually limited to the essential basics. It is for this reason that rural household energy requirements are restricted to cooking and lighting because individual families do not have the capacity to acquire suitable energy for other applications. These features clearly point at renewables and decentralized systems as the most suitable energy arrangements for the rural areas. Even if the East African countries were to extend the national grid lines to every household in the rural areas, the cost of electricity would be too high for the people to afford at the present poverty levels. Further, maintenance and service costs would also be too high for the utility companies due to large distances that would have to be covered by the technicians to reach isolated points. Consequently, the reliability of supply would be very low and this would erode the users' confidence in such a source of energy and force people to resort to other means of getting energy and most likely they would revert to the old practice. Such a trend of events cannot be generalized for all the rural areas since settlement distributions and living conditions vary from place to place. But the fact that decentralized energy systems hold the key to rural electrification still remains. The readily available energy sources such as solar, small hydro, wind and biomass are the best choices for rural application and they also offer the best decentralized supply technologies. However, it is important to first and foremost obtain accurate data on their distributions and potentials to facilitate proper identification of pioneer development and growth centers where they can be managed by small rural units such as households, villages or other organized community structures. Similarly, the ownership and management responsibilities can be placed at these levels with local village chiefs and other leadership structures given the responsibility to ensure that the facilities are properly operated and regularly serviced. In doing this, the systems are likely to be sustained if the users value the services provided. It is however important to identify suitable nucleus centers where community energy facilities can be established to initially provide highly desirable services and also demonstrate the viability of such resources. The services can thereafter be extended to the surrounding communities based on demand and existence of appropriate management structure. Such nuclei could be schools, hospitals, and organized community centers. Participation of the communities is therefore an important prerequisite for the success of these decentralized systems. However, because of the growing population and increasing demand for energy, supply of energy by these facilities should be reviewed every few years to assess their suitability and adaptation to the changing circumstances. If for a particular area the population density has increased to a level where centralized facility can be more effective then they can be replaced by introducing more suitable systems including extension of national grid power lines to the area. It is in this context that rural electrification can be expected to succeed and there are no better decentralized sources than solar , wind, small hydro, agro-based co-generation and biomass.

All these are renewable energies with both socio-economic and environmental advantages that can give the desired and sustainable impetus for rural development. There may be arguments against these energy resources but the fact still remains that under the present rural situation, they can be developed and used where the people live and with good possibility for community participation that is needed for successful application and sustainability. To prepare the ground for this success, suitable socio-political environment must be established to regulate the process so that some reasonable level of quality control and safety of applications is established. This may require the introduction of new legislations and the establishment of institutional frameworks that would specifically handle rural energy development.

All the new arrangement must also fit into the current energy initiatives such as the independent power production regulatory measures. In addition, there is greater need for effective coordination between national industrial energy needs and rural energy requirements. The two energy supply objectives cannot be achieved by a single provider without the risk of ignoring one and this has been the situation that has stifled past rural energy development strategies. The high concentration of industries in the major towns and their regular large payments to the utility companies are too attractive for a power producer to consider the cumbersome management of rural energy network. Different but well coordinated approaches should be used in order to succeed in both fronts and in this regard co-generation should be a rural renewable energy initiative that can link well with national energy supply network. Similarly, small hydro power generating facilities should be considered under rural energy initiatives but designed to link with national network as and when necessary. Some of these, depending on the situation, can be completely decentralized with localized supply networks while others would necessarily be limited to households, village, small organized communities and institution. Unlike urban energy supply systems, rural energy supply requires a lot more involvement by the users because distribution is individual-centered and not "area" focussed like the urban system. The individual consumers must therefore take some responsibilities.This is one of the reasons why some level of community ownership in a participatory sense is important for the sustainability of rural energy supply systems. Small systems like photovoltaic lighting units and biogas plants which can be owned and operated by a single household should be actively promoted by providing appropriate incentives to the users in the form of technical assistance, subsidy or tax rebates. Other incentives such as school fee waivers and subsidized medical treatment can also be considered. These are strategies that would very quickly sensitize rural communities to develop clean energy culture that would eventually move them away from reliance on raw biomass energy. The policy would also instil in people the energy conservation practices that would make significant contribution in reducing emission of greenhouse gases into the atmosphere as discussed in Chap. 6.

7.1 Human Resource Development

Most of the renewable energy technologies have been developed to a level at which they can satisfactorily serve the intended purposes in developing countries. Unfortunately for East Africa, like any other developing region, research and development of these technologies were done elsewhere without any meaningful participation at the local level. Most of them found it easy and less expensive to engage in the low level traditional biomass-based technologies such as charcoal and wood stove developments which have not made any significant impact in energy use patterns. Attempts to cover biogas and gasification conversion processes have not worked well while the use of energy from co-generation facilities was restricted by energy control regulations. This means that a lot of studies will have to be done to prepare the ground for proper and effective response to energy policy changes and technology transfer processes. Many renewable energy technologies particularly photovoltaic technology and large wind generators will have to be transferred into the region. Technologies that are not so complicated like solar water heaters, solar cookers (household and institutional), small hydro and wind machines should be developed by local personnel using locally available materials. Local skills should be developed and used in all installations and maintenance of renewable energy technologies. In order to achieve this, it is necessary to instil some energy culture in people so that the society becomes conscious of how to use and conserve energy from different sources. It is also necessary to develop the critical mass of human resources which in this context includes manpower, skills, knowledge and accumulated experience in dealing with all aspects of renewable energy technologies at all levels from craftsmanship to skilled applied scientists who can spearhead further developments and modifications suitable for local conditions. These people must be deployed in positions relevant to their training backgrounds. Placing skilled personnel in positions where they can effectively apply their professional knowledge is very crucial in any development process. This is perhaps one area where leaders of East African states have shot down renewable energy initiatives as they use political interests and ethnic considerations in appointing qualified people to positions of influence. The result of this is that a lot of people are given positions in which they do not use the professional skills acquired during the many years of training. In many instances, the morale and work ethics are eroded by appointing a relatively less qualified person as the head of more qualified and experienced colleagues. These practices are more effective in retarding technological development than any form of corruption and any wise leader must avoid them at all costs. Development in the region has severely suffered from this practice particularly in Kenya where it has been the norm in government ministries and departments. Consequently, the government receives inappropriate advices that lead to wastage of scarce resources in addressing misplaced priorities to the extent that even general repair work and routine maintenance become expensive and difficult to carry out. One example of this vice is the small hydro facilities that were introduced by colonial settlers. A good number of them in the region could not continue operating because of the so-called lack of qualified personnel to maintain them. A study carried out in Uganda confirmed this at a time when there were many highly qualified Ugandans in various branches of engineering (mechanical, electrical, civil etc) working in various government offices. Thus, human resource development in East Africa does not mean beginning to train people on renewable energy technologies today but starting with proper assessment and meaningful deployment of those already available, and they are many. Since the 1970s quite a significant number of many East Africans have been trained on various types of renewable energy technologies and therefore a very effective task force for the development of renewable energy technologies can be very quickly constituted. In addition to this, it is important to set up at least one center of excellence in research and development of renewable energy technologies. The center should be primarily charged with the responsibility of carrying out research and development work on renewable energies but should also be in a position to provide data and investment guidelines on renewable energies. Some suggestions on how such a center can be established with the prevailing limited resources are discussed in Chap. 8.

To build the capacity of the center to an effective productive level, there would be need to fine-tune existing renewable energy technology courses in institutions of higher learning in the region. Renewable energy technologies have been fairly well developed and therefore the center should not waste its resources trying to re-invent the wheel but should focus on resource identification and technology adaptation to suit local socio-economic and resource accessibility circumstances. This task is important and enormous and should not be left to the leadership of the center alone to decide. The activities must address national energy development goals as set out in national development plans and this means that it should have adequate annual budget allocation in which disbursement of funds is based on the progress made in achieving the set goals. In this regard contract research and technology development whose achievements are easy to measure should form the basis of the activities of the center and government ministries and departments should be the main sources of these contracts. These activities should be linked with those of the private sector particularly non-governmental organizations at the dissemination stages. All these will call for the establishment of suitable regulatory framework to ensure that procedures and requirements are followed and fulfilled and that adequate resources in both manpower and facilities are made available for this. The efforts of many non-governmental organizations in the 1980s and 1990s on the development of renewable energy technologies would today have made a greater impact in the region than they did if they had been given organized institutional backing beyond mere lukewarm non-interference attitude. Theoretical policy changes that are prominently presented in official documents without the necessary structures on the ground will not lead to the solutions of rural energy supply problems. There is need to do more in the form of institutional arrangements that would effectively link the demand side with supply strategies and use decentralized renewable energy programmes as the principal means to satisfy the requirements. Conditions for environmental protection could be used as indicators for choosing the method and source for energy production so that independent power suppliers are guided accordingly. This is not a new concept as it has been used by many developed nations to encourage independent power producers to develop solar and wind energy systems. For example, Germany, which is one of the world's leading nations in the use of wind energy is also very conscious and strict on environmental protection laws and regulations and uses these principles through appropriate legislation and institutional arrangements, to encourage independent power producers to use renewable energy sources. A similar institutional approach in East Africa would not only encourage independent power producers to turn to renewable energies but would also encourage more active local participation in the development of these technologies. Existing energy related organizations should be strengthened and used as energy development focal points. In East Africa, a regional approach to energy issues preferably with coordination of the reconstituted East African Community and the newly established East African Power Pool should be used to improve the region's energy infrastructure. Again, these institutions would require suitably qualified personnel who can achieve the objectives. The linkages and experiences gained by non-governmental organizations can serve as useful tools for better networking and information exchange on renewable energy technologies. When compiling an inventory for renewable energy resources, it is important to include detailed information on institutions and agencies involved in energy development issues such as training, dissemination, energy technologies, energy policy studies as well as their potential and socio-cultural constraints. Such information would encourage exchange of experiences, research collaboration, dissemination, and training capabilities on a regional basis so that larger markets and standardization procedures are developed. This is expected to attract both local and foreign investors and facilitate cost-effective exploitation of the region's huge renewable energy resources.

Difficulties in financing renewable energy initiatives have been a major inhibiting factor in the development and application of renewable energies. It is therefore necessary to establish easy and flexible financing mechanisms specifically for renewable energy technologies. There are a number of ways in which this could be achieved but it would be important for such structures to include credit schemes, higher taxes on fossil fuels, and attractive tax rebates or wavers on renewable energies. It is also worth considering the establishment of national fund for renewable energy developments. There has been, in East Africa, fuel tax, which is paid at the petrol or gas stations whenever one buys fuel. A good fraction of this money should go into the development of local energy resources so that the tax is justified by using it in a way that enhances patriotism. This could be done through contributions to the renewable energy fund or by putting it into a revolving fund to facilitate easy and flexible financing of renewable energy programmes including support to sympathetic credit institutions. The process of administering such funds involves accurate assessment of the need and circumstances of the applicants and therefore would be more effectively handled by local institutions that understand both cultural and social factors that have implications on ability to respond to loan repayment obligation. Financing arrangements are very sensitive matters that, if not well managed, can kill even the best of the initiatives and so the regulatory mechanisms for controlling this aspect must be effective and just. Obviously there would be need to establish or reorganize existing institutions that would devote adequate time and resources, in a coordinated framework, in order to satisfy both renewable energy developers and users. It is through all-round coordinated and concerted efforts that the success of renewable energy technologies can be realized in East Africa. The challenge is therefore focusing on three major areas: (1) formulation of suitable renewable energy policies that would encourage the development of renewable energy technologies; (2) establishment of institutional framework for effective monitoring and coordination of renewable energy activities including the enforcement of quality control measures; and (3) constitution of easy and flexible financing mechanisms to provide the impetus required in the whole process. These should be viewed and accomplished in the background of well-documented data on all aspects of renewable energies including national and regional distributions, economic viabilities, potentials and possibilities of modular developments. It is expected that the local human resources would be developed in all these areas and effectively deployed appropriately. Many of theses issues have not been properly addressed because there is not enough skilled manpower that could formulate suitable and implementable policies for the rural conditions. Human resource development must therefore be seen in its entirety to ensure that every initiative gets the desired support from all relevant sectors.

7.2 Biomass: The Challenges

Although the authorities in East Africa have not given biomass the recognition it deserves, it is however widely accepted that it plays an important role in the developing countries in general and that its use is not by choice but a survival means for the poor rural communities. It is a reliable source of energy that does not have to be stockpiled in large quantities as it can be obtained on a daily basis. In spite of its importance, biomass has generally been ignored in official energy planning strategies and yet its use is associated with environmental problems, whose impacts affect the whole nation. Furthermore, it can be depleted if its regeneration is not carefully planned and this would have far reaching consequences on the environment such as soil degradation and changes in weather conditions that would have adverse effects on agriculture. There have been a number of activities on agro-forestry and tree planting in general but not specifically for energy purposes and in many cases the ministries of energy in the region were not actively involved. The evidence that biomass use far exceeds its natural regeneration is common knowledge and so is the undesirable impact of its careless and unplanned use. Biomass is also used for construction and it is also destroyed to give way for new settlements and agricultural land. All these activities are increasing the demand for it while its supply is diminishing. This is definitely a very dangerous trend. In addition, its use is a great health hazard to the women and children who use it for cooking. Burning of biomass, particularly firewood and charcoal, is associated with emissions of carbon dioxide and the toxic carbon monoxide. Users of these fuels are therefore regularly exposed to the risks of developing diseases that affect respiratory system.

The common practice when these fuels are used indoors is to simply let the smoke find its own way out of the room. There is virtually no attempt to vent the smoke out of the room. Indoor air pollution from smoke is suspected to be the major cause of respiratory diseases that kill a large number of women and children in East Africa. In addition to this, the emitted carbon dioxide and carbon monoxide are well known greenhouse gases that cause undesirable increase of global warming. The other concern regarding the use of biomass is the rate at which deforestation and land degradation is taking place. In recent years, the escalating oil and electricity prices and their intermittent supply in many areas encouraged many institutions to turn to wood and charcoal. In Kenya alone, more than two million tonnes of fuel wood is consumed per year. This is based on a modest estimate of about two kilograms of fuel wood per household per day. Large institutions like hospitals and colleges use between 5 and 10 tonnes of fuel wood per month and this raises the estimated figure by more than 100,000 tonnes per year further increasing the total annual fuel wood consumption in the country. Regeneration is difficult to accurately assess, but it is obvious that it does not match consumption rate and therefore there will be an imminent fuel wood crisis if nothing is done to slow down consumption rate and to intensify regeneration activities. It is generally accepted that, for the purposes of comparing consumption rate against existing wood reserve, ten tonnes of fuel wood is equivalent to clearing one hectare of forest, but this depends on the forest density, which is very low in most parts of East Africa. The unfortunate situation is that the already visible land degradation and deforestation in the region have not been given adequate attention from the energy point of view. They are normally assumed to be the responsibilities of the ministries in charge of agriculture and natural resources while energy-related health hazards associated with fuel wood use are handled by the ministry of health. Under this arrangement, it is very difficult to implement preventive measures because energy consumers will continue with their practices as if these matters do not concern them.

There are, however, a number of options in addressing the problems associated with biomass energy use. Developing an energy-sensitive culture in which energy conservation and application risks are well understood by the society is the foundation on which all energy issues should be based. Alongside this development should be the introduction of efficient methods of converting energy from one form to another especially biomass conversion processes as well as promoting use of efficient stoves. Improved charcoal kilns should be more aggressively pursued and once the technology is widely understood, the use of inefficient techniques should be banned. But first, the organization of charcoal production must be improved to a level where use of these improved conversion technologies can be effectively practiced. Tested and successful kilns such as the Katugo and Mark V in Uganda and the oil drum in Kenya should be carefully re-examined and their improved versions promoted in East Africa. The stoves that use charcoal and wood should be comprehensively re-analysed with a view to promoting efficient models. Special attention should be given to those stoves that have had a measure of success in the region such as the Kenya Ceramic Jiko (KCJ) and the maendeleo wood stoves. These stoves have proved that they have significant desirable advantages such as less consumption of fuel and achievement of more complete combustion of fuel than the traditional stoves. Similarly, improved institutional stoves should be seriously considered at national levels and the governments should assist in their promotion, dissemination and use, particularly in government rural institutions. At the same time, the governments should support efforts to develop stoves and devices that can use other types of energies especially solar and fuel briquettes. In this regard, use of biomass fuels should in future be restricted to high temperature heat requirements which cannot be achieved through alternative rural energy sources. Low temperature needs such as water and space heating should be achieved by other means and not from biomass fuels.

At a larger industrial scale, biomass is an important energy for co-generation and also for biogas and producer gas production. Of these applications, only producer gas may require wood. The other processes make use of animal wastes and agricultural residues and so the risk of running short of supply is very low. The cows, chicken and other animals will always make their wastes available for a biogas plant while for co-generation by sugar millers, bagasse is the main waste and is always available in large enough quantities to generate electricity for the industry and the surrounding communities. The full regeneration cycle of sugarcane, which produces bagasse, is between 15 and 20 months and given that the milling capacity of the factory is known, it is possible to determine the size of the sugarcane farm that can adequately and continuously serve the factory for both sugar milling and co-generation activities. Sugar milling enterprises are some of the agro-based industries with the lowest failure risks and this has been demonstrated in East Africa where despite civil wars and adverse direct political interference that forced many industries to close down, sugar millers survived. The worst hit were in Uganda and Kenya where they were mismanaged and plundered but they still went through all these successfully - a proof that these are the types of industries that are suitable for the unpredictable African political conditions. With the additional economic empowerment in co-generation and role as independent power producers, these establishments are sure to create a significant economic impact in the rural areas and enable the communities to broaden their energy options beyond biomass. At least the few rural elite will quickly move from biomass to electricity, LPG and kerosene for cooking. The improvement of the economic status of the rural people will enable them to acquire the capability of switching to cleaner modern energy technologies and increase artisanal job opportunities. The economies of East African states depend mainly on agricultural outputs of coffee, sugarcane, tea, pyrethrum, cloves and cereals that double as staple foods as well. Co-generation is an industrial activity that is economically viable for agro-based industries especially sugar processing industries, which have large amounts of industrial waste in the form of sugar cane bagasse. Thus co-generation is a rural-based industrial undertaking that has the highest potential for improving rural electrification. Its advantages include a wide range of choices regarding the location of the factory that automatically leads to the spontaneous development of new human settlements and vibrant trading centers that create income generating opportunities for the local people. Such new settlements can serve as the nucleation centers for rural electrification. With the two industrial outputs of the processed goods and electricity, the companies are sure to make profits at lower prices of their products and the factory need not be too large. Local investors would therefore be encouraged to establish medium or even small-scale agro-processing industries with commercial power generation components. The diversification of electricity generation using various renewable energy sources would create opportunities for broad-based rural energy development programmes.

Production of producer gas at commercial level mainly for decentralized rural electricity generation for localized power supply to villages and market centers will also improve rural economic status of the people and uplift their lifestyle. Independent power producers particularly indigenous groups or companies should be encouraged to participate in the development and promotion of such enterprises. Producer gas technology requires too much attention and specialized knowledge to succeed at household level and so focal units should be, among others, organized villages and market centers where the economic scale of the system is viable. The best focal units would be schools, hospitals and any other organized rural entities. Biogas, on the other hand, is a household energy technology where its management and use are a family concern. The possibility of mismanagement and irresponsible use increase with the number of people using a common unit and therefore the responsibility of managing and using biogas plant should be kept at a family level. Statistical data collected in East Africa indicate that more communal biogas plants are generally poorly managed than those owned by individual families. Biogas plants managed by institutions have also performed relatively better than the communal ones. There are also indications that portable biogas that can be used by those who do not have their own plants would be more attractive and would enable the governments to reduce importation of liquefied petroleum gas (LPG) in the region. Technologies that would make this possible should form part of the energy training programmes in institutions of higher learning particularly those institutions which are already engaged in various aspects of animal production such as Sokoine University of Agriculture in Tanzania and Egerton University in Kenya. Other government institutions, which keep animal farms and poultry, should be the first to establish biogas plants for both demonstration and normal application purposes. Waste treatment lagoons in municipalities, rural industrial establishments, hospitals etc would be more useful if modified to act as biogas plants to provide energy to those living close to them. Such biogas production methods have been successfully applied in many parts of the world including the United States of America. One advantage of doing this is that it also helps in ridding the surrounding of the strong smell from these usually exposed lagoons. The actions on the waste by microorganisms that participate in the production of the gas results into final stabilized slurry with soft smell and good soil conditioning qualities.

Socio-cultural bottlenecks regarding the use of human waste should be identified and appropriate measures put in place to overcome them. Another aspect of biomass energy, which requires considerable attention, is the production of fuel briquettes from agricultural residues and other biomass wastes. Although the technology of producing fuel briquettes is known and has been used in the region, there has not been a serious attempt to develop special stoves that can efficiently use this form of fuel. A number of people have used ordinary wood or charcoal stoves to burn fuel briquettes while such stoves are not suitable for briquettes and so their poor performance discouraged many users. The development of fuel briquette stoves is a task that should be given more attention by energy specialists and research institutions. The absence of such stoves has hampered the wide scale use of the already existing fuel briquette machines. This is just one good example of how poorly renewable energy development has been coordinated in the region because briquette machines should have been introduced together with some prototype stoves but it appears that the machines were promoted in isolation and so people did not understand their importance. At some stage, some unfocussed proposals were made about the stove but there was no concrete research and development work done on it and the idea faded away as quietly as it came. There is need to formulate an integrated approach in the development of renewable energy strategies for East Africa in order to find the right mixture of approaches that would produce the best result.

It is a fact that the position of biomass as an important energy source will continue for many years to come and therefore ways of efficient use, conservation and regeneration should be considered and indeed included in all rural energy planning processes. It is also true that there are a variety of biomass wastes that are not being used as energy sources. Their application should be developed to an economically attractive level for the rural situations where they are abundantly available. Some of these wastes are regarded as environmental pollutants when they could be used to provide the much-needed heat energy. While advocating for improved and attractive use of biomass fuels, it should always be remembered that the consumption of biomass is associated with emissions of carbon dioxide, which is one of the most effective greenhouse gases. Therefore all biomass applications should give due considerations to the environmental impacts of using it as fuel and appropriate measures incorporated to address this. The situation, at present and in the foreseeable future, is that the use of biomass as a major energy source for the rural communities will continue and therefore it is unthinkable to neglect it in the planning and development of energy resources. East African energy policy makers and planners must directly address it and endeavour to put in place institutional, financial and consumption strategies that can effectively and efficiently guide its sustainable use. To achieve this, it will be necessary to develop and promote other renewable energy resources in order to ease the pressure on biomass. In this regard, it is important to note that much of the biomass energy is for heat requirements and not electricity or lighting and so the efforts to ease the pressure on biomass should first and foremost consider rurally affordable renewable heat sources. If this heat is to be obtained from electricity, it must be almost as cheap as biomass fuel and the payment terms must be easy and based on consumption and not anything else that would raise the cost. The other alternative is to improve the economic status of the rural population to a level where they can afford more expensive energies. This is a task that cannot be achieved overnight and will take several decades. Whatever the choice, biomass regeneration efforts must be intensified and addressed through appropriate programmes that would promote reforestation and agroforestry activities. This means that nationwide culture of tree planting should be revitalized and balanced against other land-use patterns especially those that arise due to increased population pressure. Experience gained in previous tree planting programmes should be used in making the right choices of tree species that would not interfere with other vital land use activities such as agriculture and livestock production. Two other aspects of biomass energy that might in the near future play key roles in the transport sector are power alcohol and bio-diesel. Global petroleum politics is getting more and more interesting and is apparently the source of modern international conflicts. Nations are already getting concerned about access to oil products as global reserves continue to decrease and while the demand is going up. This situation should be of grave concern to the developing countries, which have neither the gun nor the money to secure oil supply. The survival of such countries may well be determined by their own ability to use appropriate technologies to convert biomass materials into liquid fuels that can be used in industrial and transport sectors. In East Africa, Kenya has had some experience in power alcohol production and should strive to remove the bottlenecks associated with its production in order to make it an economically viable energy source. As for bio-diesel, East Africa has no experience with its use or production but the potential for producing bio-diesel crops exist. This will however require adequate investment in relevant research and development to establish the characteristics of suitable crops and efficient oil extraction methods.

7.3 Solar Energy Applications: The Constraints

In general, household energy requirements, rural or urban, are for high temperature heat and lighting. Household appliances and home entertainment are the pleasures of those who can afford energy beyond heating and lighting needs. Home entertainment however, has more energy source options (in the form of dry cells and lead acid batteries) than lighting or heating. Both these needs can be met through photovoltaic and thermal conversions of solar energy into electricity and heat respectively. These two processes can be achieved independently using different specific techniques and this is a characteristic of sunlight that makes it an important energy resource for the rural situation in which two different sources of energy (wood and kerosene) are normally used for cooking and lighting. In deed a few rural families are already using solar energy for lighting and there is also another small number using solar cookers. The biggest constraint in the application of solar energy is that it is available only during daytime and yet energy requirement is more critical during the night. Suitable deep discharge batteries are readily available but they substantially increase the cost of solar photovoltaic systems because they are used together with additional system components such as charge controllers. The applications of photovoltaic solar systems are therefore limited to middle and high income groups who can afford to buy them. One 50W panel which can provide lights for a family of 4 to 6 people living in a four-roomed house costs about US\$ 300 which comes to about US\$ 600 together with the basic components like battery, charge controller, light bulbs, cables and the necessary circuit components. Over 90% of the rural population in East Africa, have an average monthly income of less than US\$ 50 and cannot afford to buy the system unless some easy and flexible payment arrangements are provided. The region lacks suitable fnancing mechanisms such as hire purchase, loans and subsidies through which the low income groups can afford to purchase solar lighting systems. A few financal institutions that have provided credit facilities were faced with problems and costs of tracing defaulters and had to withdraw the service. Existing financial institutions therefore are not willing to get involved in short or long term financial arrangements with individuals whose occupations are not clearly defined. The problem is also compounded by the fact that the majority of the rural population do not have legally bound and credible employers through whom they can repay loans and credits.

Another discouraging factor is the instability and irregularity of income for most people in the rural areas due to the temporary nature of their employments. Rural establishments such as sugar factories, hospitals, and schools take advantage of the high rate of rural unemployment to engage workers on temporary and casual terms with very low wages. The low income and unstable employment situation make it very difficult to find slow-moving items in the local market particularly the more expensive and non essential goods. Consequently, renewable energy devices such as solar photovoltaic panels are not readily available in the rural markets. Most PV stockists are operating in large cities and towns and so are far removed from the typical rural population, and in deed, many of the people who are already using PV systems are those who are fairly well informed or whose relatives are familiar with the technology. So there are two factors that inhibit the use of solar PV systems. One is the price which many people cannot afford under single payment terms even though, in the long run, it turns out to be almost comparable to other sources that are considered to be cheap. The second one is lack of information on how to correctly use a PV system within its own technical limits. A number of people have been disappointed by PV systems because of incorrect use that overloads the system to the extent that power is exhausted too fast making it unlikely to recharge to full capacity the following day. Often the system is also used during daytime and so denying the battery the energy for full recharge. A system thus used would not satisfy the demands of the user. It is therefore necessary to educate people to understand that a PV system cannot be used in the same way national grid electricity is used. People must understand the various reasons for energy conservation particularly the prudent use of renewable energies. The location of supply points and unavailability of technically qualified personnel to give the correct information on application also, to some extent, limits the use of PV systems. The various non-governmental organizations that promoted renewable energy technologies in the region considered PV systems as a purely business concern and therefore did not carry out any organized promotional activities on the systems. The commercial PV dealers who were expected to promote their use also did nothing to let people know about them. Incidentally, even international development agencies that support local non-governmental organizations were and continue to be very hesitant in supporting the dissemination of PV systems. They instead chose to support wood energy technologies especially those that focused on charcoal and wood stove developments. The logical policy would have been to support viable initiatives that would reduce the pressure on biomass and fossil fuels and solar energy provides the best opportunity for this. Admittedly, the marketing of PV systems in the region is reasonably developed but they are considered as slow moving items and therefore are not stocked in large quantities. Given the past experience and the knowledge on existing drawbacks in various energy systems, non-governmental organization are now well placed to conduct effective information dissemination on PV systems while at the same time setting up demonstration units in strategic centers such as schools, rural grocery stores etc. It is not expected that the price of PV systems will drop to affordable levels for the rural peoples in the near future and therefore deliberate efforts must be made to establish infrastructure for easy and flexible financing systems as the basis for widespread application of PV systems. This is an important prerequisite for the success of PV systems applications in the rural areas and it demands going down to the bottom of the community to deal directly with individual users. Cooperative societies would be in a better position to provide suitable financing programmes but most of them have failed to meet the expectations of members and are struggling to survive. Furthermore, memberships of cooperative societies are open to those who have regular monthly incomes and whose employers can manage the check-off loan repayment system for the society. The majority of the rural population do not have regular incomes and are generally engaged in some sort of vague and weak self-employment - a condition that is not catered for by cooperative societies. Above the community level, national governments have not given solar energy sufficient support to compete favourably with other energy resources. Great attention and huge foreign currency and budgetary allocations are effectively given to oil and large centralized power generation facilities that benefit less than 25% of the total population and almost nothing is given to the development of renewable energy technologies that would benefit everybody. The examples of this are abound not just in East Africa but also in many developing countries. For economic reasons, the action may be right but it needs to be balanced with the people's energy supply aspirations. This is a remarkably great contrast to what is happening in the developed economies where there are wider energy options but still they put a lot of emphasis on renewable energies. In these countries, significant support is given to both research and application of renewable energy technologies. The support includes tax rebates, attractive energy buying price from independent power producers using renewable resources, which in effect means subsidizing renewable energies, and significant renewable energy research grants to research institutions including universities. There are also scholarships available for those who wish to study renewable energy technologies at postgraduate levels. These are some of the approaches that should be considered in East Africa in addition to institutional arrangements. It is not right to import solar energy conversion devices when the capacity can be developed to locally produce them. East Africa receives more solar radiation than those countries that supply PV systems to the region and it is possible to correct this anomaly simply by shifting the emphasis to renewable energies.

As mentioned earlier, solar energy has another attractive aspect, that is, it can be converted directly into heat, which is another form of energy in great demand. There are two levels at which this can be applied: the low and high temperature applications. Low temperature applications include water heating, space heating, and drying of agricultural products in order to prolong their storage life. The technologies for these applications are very simple and quite a good number of them have been developed in East Africa and are in use. However, low temperature applications are not priority requirements as people can do without such heating devices by using open air sun drying. The climatic conditions in the region do not demand regular space or water heating except in a few highland regions where night temperatures can go down as low as $10°C$. But such areas normally have abundant woody biomass for general heating purposes. Furthermore, water heaters are not suitable for rural use because they require pressurized water supply source that can force water to flow through the heater located on the roof of the building. More than 90% of the rural houses are not supplied with pressurized piped water. People normally fetch water from nearby streams, rivers, shallow wells, or springs using portable water containers. Most solar water heaters that are in use in the region are installed in the cities and large towns where pressurized piped water facilities are available. Therefore the fact that these technologies are not used widely in the region is not a cause for concern since the rural living circumstances are not suitable for their application. The second level is the high temperature application, which is mainly for cooking and water boiling. Most water sources are contaminated and therefore drinking water must be boiled to avoid water-born diseases. These are the two main areas that consume a lot of biomass energy in the rural areas and so the use of solar energy to meet these needs would be a great relief on biomass. Solar cookers have been developed in the region and various designs are in use. However, the majority of the people have not been sensitized enough to appreciate this source of energy and exiting technologies for harnessing it. The available solar cookers are not as expensive as other solar devices and therefore financing should not be an issue since most rural people can afford to acquire them. So far solar cooker production centers are very few and demonstrations are usually done in scattered isolated places so that most people have remained unaware of this alternative way of cooking. The second constraint is that cooking during daytime is an activity that people would like to conduct quickly and at a particular interval of time between more pressing daily responsibilities. Night cooking is also done quickly in the early part of the night to save on energy for lighting and also to have more time to rest and therefore a fast-cooking device is desirable. Solar cookers are generally slow and the availability of direct sunlight when needed is not assured. Night application of solar heat is currently not viable since effective high temperature storage systems are lacking. These characteristics and lack of energy control have inhibited their wide scale use in the region. In addition, using them require too much attention as they have to be regularly moved to face the sun all the time in order to constantly receive maximum direct sunlight. Some of their designs such as the more efficient ones that use parabolic reflectors to concentrate sunlight on to the pot are not convenient to use, particularly for preparing some traditional foods. There is therefore need to carry out more educational and promotional activities on them and also encourage further research on user-friendly designs and materials with high thermal conversion efficiencies. These improvements should be conducted alongside training programmes that would enable existing local stove manufacturers to include solar cookers as one of their products. Thus there is still a lot that needs to be done on solar cookers before they can make any significant impact on rural energy scene. In general, for the conversion of solar radiation into high temperature heat the challenge is to come up with suitable designs that can meet the traditional cooking practices. The technology itself is simple and has proved that the devices can achieve sufficient high temperatures so there is no problem with the technology itself. On the other hand, conversion of sunlight into electricity is more complicated since it requires the development of suitable semiconductor materials that can be used to make solar cells. The solar cell manufacturing technique, from raw material processing to actual cell production requires specialized knowledge and use of high precision equipment. With committed support from the national governments, the region can develop the required technical skills and acquire the equipment to produce solar cells locally.

The situation outlined above clearly indicates that solar energy constraints fall into the following four areas:

(a) Awareness: Many people in East Africa including policy makers are not aware of the potential benefits of solar energy systems and so there has neither been any attempt to provide extension services nor promotion campaigns in the public sector. All these have been relegated to the private sector, which is more concerned about making profit than spending on awareness campaigns.

(b) Affordability: Acquisition of solar systems need large capital that most people in the rural areas do not readily possess and there is no financing mechanisms in the rural areas for it.

(c) Standard of solar system components: Some solar energy components available in the market are of low quality due to lack of enforceable standards. This, in addition to their high costs, has discouraged potential users.

(d) Technical capability: Poor installation and maintenance done by untrained technicians result into malfunctions that lead to disillusionment and loss of confidence in the systems.

Solar energy promotion in the region must consider ways of removing these constraints.

7.4 Wind Energy Challenges

Wind energy like hydropower is not easy to efficiently convert into heat unless it is first converted into electricity through a generator. But it is quite suitable for performing mechanical work such as water pumping or grain grinding. Although wind is available everywhere, its strength varies with time and from location to location. The success of its exploitation is therefore highly site specific so that the study of wind conditions for the site must be carried out over a long period of time before a suitable machine can be installed. In addition to this the machine must be designed for the intended function. For water pumping, high torque is required to turn the blades so that the pump receives enough energy to move through all its mechanical linkages. If the machine is to be used for electricity generation then high-speed machines are required. Thus it is important to match wind turbine to the prevailing wind regime at the selected site. Failure to do this can result into inefficient operation of the machine. The diurnal and seasonal variations of wind speeds at a given location also mean that the machine will not operate all the time. Electricity from a wind generator without storage facility is therefore unreliable but wind is still a better alternative than solar because it can operate day and night. For water pumping, the intermittent operation may not be a serious drawback since the water is pumped to a storage tank from where it is drawn only when needed. However, knowledge of wind regimes, regular updates of this information and its availability to both policy makers and energy experts in the whole region are important prerequisites for wind energy development. Wind pump technology is relatively simpler than wind generator and there is adequate local expertise in East Africa to produce and manage them. The problem at the moment is that the demand for wind pumps is very low and therefore the manufacturers tend to produce them on order. This is partly due to the high cost of the machines and partly due to limited use of ground water, the long distance from water points and land ownership issues. The use of wind generators is even more limited and there are no well established local manufacturers and so there is need to aggressively develop the capacity to design, construct and install wind generators of different sizes and for various applications.

The polytechnics and institutes of technologies should have more practical approach to the development of appropriate energy devices particularly wind, solar and hydropower conversion machines. People living in the developing countries quickly get into new ideas by copying examples and so these institutions must lead by producing and using these devices for people to develop confidence in them. If necessary, an authority should be established to ensure that these technologies are developed and, first and foremost, used by the same institutions. This is one way of developing new technologies and promoting their application at the same time and it goes down well with the African customs, which believe that a visitor would be easily encouraged to eat your food if you are also eating it. The present institutes of technologies and polytechnics should be strengthened and refocused to effectively pursue their original mandates, which included design and fabrication of energy devices. In this regard, their capacities with respect to qualified staff and well-equipped workshops should be the first priority in all technical institutions in East Africa. To provide incentives and attract dynamic students to these institutions, it may be necessary to expand opportunities for skill development so that technically smart students can develop their capacities beyond artisanal level. The need to do this was recognized many years ago and in deed the establishment of universities of technologies was proposed but unfortunately, regional and ethnic politics allowed such universities to revert back to normal traditional university disciplines and completely abandoned the original mandate. Thus, proper management of and commitment to relevant educational and training system would enable the region to adequately address technology development issues so that such simple energy technologies like wind turbines, hydro turbines, solar energy conversion and other related devices can be locally produced and promoted. Thus capacity building for home-grown technologies is the immediate step that should be taken by East African countries if any significant step is to be made in technology development in general. It is through human development for renewable energy technologies that technicians, technologists and engineers can be produced to sustain any desired renewable energy programmes. Many of the renewable energy projects in East Africa were managed by expatriates through donor-funded programmes and most of them could not be maintained as soon as the experts left. This is an expensive experience that should be avoided in any future renewable energy programmes.

Medium and high income groups owning properties in the rural areas should also be encouraged to use renewable energy technologies even though they can afford the readily available but equally expensive energy resources. But this can only succeed if security of supply is assured and the supply is reasonable priced to attract the consumers. In this regard it is important to address the following constraints facing dissemination of wind energy technologies:

- High capital cost that makes wind energy less attractive compared to oil-powered generators;
- Inadequate data on wind regimes in the region;
- Lack of awareness about the environmental and economic benefits of wind technology;
- Lack of enforceable standards to guarantee quality;
- Lack of user friendly credit schemes and financing mechanisms.
- Limited after-sales service;

7.5 Small Hydropower Systems

Large hydropower potentials in East Africa have not been fully exploited because such developments are proving to be more and more expensive for the regional economies to absorb. For example, in 2005 the government of Kenya had to stop the development of the 77 MW Arror hydropower station because the project could not meet the least cost development criteria. The cost of electricity generated from that project, if implemented, would be higher than the present cost of electricity from other hydropower facilities. The unfortunate thing is that at the time this decision was made, the government had already spent some substantial amount of money on it. Such wastage of limited financial resources can be avoided if power development policies switched from large to small hydroelectric power development. East Africa has many permanent rivers whose volume flow rates are not sufficient for large hydropower stations but on which small hydroelectricity facilities can be established at affordable costs. Many sites suitable for this purpose have been identified and there are many others, fairly well distributed in the region, for which feasibility studies have not been done. The studies that have been done on the known sites were conducted by NGOs, environmentalists and individual researchers but the governments that are expected to have detailed information on these for prospective investors have not given this option some serious considerations. It is true that large and medium hydropower stations have given the region most of her electric energy - an energy form that is undoubtedly clean. However, times are changing and there are new issues that must be considered in developing more of such technologies.

The potential still exists but, as has been demonstrated by the Kenyan example and the Bujagali conflict in Uganda, the construction of large hydro dams for electricity generation is too expensive and involves a whole series of environmental questions. Some of these environmental considerations recognize the need to preserve some of these rivers for ecological, cultural and biodiversity reasons. Obviously the alternatives to large hydropower technologies are many but most of them too have their environmental drawbacks and therefore energy technologies that can be appropriately scaled to avoid adverse impacts on the environment must be considered. So it would be unreasonable to argue for total exclusion of hydropower development because there are hydro technologies that can be used within reasonable environmental and ecological guidelines and small hydropower systems provide the best opportunity for this. Although the management of large state controlled hydro schemes suppressed the real demand for electricity through unfriendly policies, they have shown that they are not viable businesses as evidenced by their poor performance resulting into huge losses. Recent reforms are expected to correct this but there are indications that total privatization of many of these utilities may not be possible since there are national security issues to be considered. It is therefore unlikely that opportunity for private sector in East African electricity supply will go beyond the present constraints. The only ray of light in this situation is seen in the development of mini hydro schemes. Better still, there is a wide range of choices of small hydro technologies that can suit different river flow regimes. Some of these include low head, high head and free flow small hydro machines that can be used with minimal environmental impact. They also offer great advantages such as flexibility, potential to keep electricity cost low and socio-cultural acceptability. In addition, they are cheap to install and manage and can sufficiently meet the local needs within a simple distribution system. The immediate priority is to identify locations of small hydro potential sites that are suitably close to active rural agro-based commercial enterprises such as coffee and tea processing centers, grain and milk collection points, and fish-landing beaches. Schools, health centers and rural administrative centers should also benefit from such small hydropower systems. The major costs of small hydro machines are associated with the equipment itself particularly if imported. Every effort should be made to reduce the cost of the equipment and this can only be done if they are made locally. Local engineering farms, research and development institutions should be encouraged to produce these machines by giving them appropriate incentives in the form of tax rebates, financial support for design and development costs and local market guarantee. Besides technical considerations, it is necessary, in general, to consider a broad range of issues in order to develop and successfully manage hydropower systems. Some information that would form the basis of these considerations are: need for power in an area; existing methods for meeting those needs; suitable energy mix for the prevailing circumstances of the people; environmental impacts of present and future methods of energy supply; the economics of possible energy choices; and priorities in technology and resource development. All these would require good planning and effective implementation strategies and not "Do It Yourself" attitude with which the authorities have treated rural energy supply in East Africa. The Viet Nam experience in which it initially relied on imports from China and later learned to produce its own similar Pico hydro machines should be a good lesson for East Africa. There are good opportunities in Uganda, for example, where less than 5% of the mini hydro potential has been developed. The factors for these opportunities include high cost of electricity from diesel and petrol generators due to high fuel prices; small hydro schemes are supported by local environmentalists; basic manpower is already available within local utility companies; low voltage mini hydro schemes favour rural conditions; and there is a fast-growing electricity demand. The case of Nebbe hospital in Uganda should serve as a good example of how bad energy policy has affected the development of a conveniently located mini hydro site. The hospital and township require reliable electricity for water pumping, refrigeration, theatre work and lighting. There is a suitable mini hydro site close to Nebbe with a capacity to supply 40kW, which would be enough for Nebbe hospital and the township but there are neither community-based development plans nor any suitable policies to encourage commercial use of such facilities in a localized framework. For years, Nebbe spent large amounts of money running a generator that could only be operated for a few hours in order to keep the cost down while an un-developed cheap reliable alternative is in the vicinity. Nebbe's case is not a unique one since there are a number of such un-utilized opportunities in the East African region.

7.6 Municipal Waste

For obvious reasons, clean energy demand in the urban centers is much higher than it is in the rural areas and is usually more than can be met by existing facilities. The generation of waste is also very intense in the urban areas due to high population density, intensive commercial and industrial activities. Managing this waste has been one of the major problems faced by practically all municipal authorities in the region. Its collection and disposal is not only a net revenue sink but is also too expensive. If properly managed, the waste can be turned into a useful resource that can generate some revenue for these municipalities. This would create the means and the need to improve sanitary facilities, which, in turn would reduce incidences of diseases that arise from filthy living conditions. With appropriate management of municipal solid waste, it is possible to generate electricity using technologies that are already available in some developed countries. Instead of requesting international development agencies to assist in developing large hydro power plants, national governments should seek assistance to develop power generation facilities using municipal wastes, as this will improve both power supply and sanitary conditions for the poor urban population. The use of waste for electricity generation is a unique solution to both waste management and revenue problems. The challenge, however, is to improve waste collection and sorting in order to separate energy waste from the rest. There is also need to identify suitable energy generation technologies that would use municipal waste as feedstock. Another energy production possibility from municipal waste is to design waste water treatment lagoons to produce biogas. This approach is different from the use of solid waste to generate electricity but energy linkages can be developed so that, although they operate as independent units, they supplement one another, if necessary. Regional authorities must begin to think of new technologies of turning waste into useful resources. This will go along way in improving the environment in the crowded residential urban estates.

7.7 Policy Issues

All the three East African governments of Kenya, Tanzania and Uganda are primarily concerned with the development and management of commercial energies: electricity and Petroleum fuels. However, in terms of energy accessibility, less than 10% of the total population in the region benefits directly from these energy resources. The rest of the people, mostly living in the rural areas, have to find and manage their own energy resources and the reasons for this go beyond accessibility. There are the questions of poverty, temporary and poor quality of shelters and the lengthy process of preparing local traditional foods. Due to a number of constraints associated with these conditions, the people have no choice but to use the available energy resources that are easily accessible and affordable within their territory and only biomass-based fuels particularly firewood fit into this category of energy, making it the natural choice. It is considered as a non-commercial energy source since it is largely collected from the neighbourhoods free of any charge. All the three states of East Africa have broad energy policies, which give guidelines on how both commercial and non-commercial energies should be managed to ensure that energy is available to everyone and that suitable arrangements are made to guarantee sustainability of all vital energy resources. On the question of accessibility, the policies seek not only to increase access to a greater number of people but also to improve services and efficiencies of energy technologies in order to protect the environment and reduce health hazards associated with energy use. This means that despite the wide gap of income levels of the people, everyone should have access to improved household energy services and efficient energy technologies. These policies have good intentions but unfortunately their implementation and hence achievement of the objectives come with costs that the people are expected to meet. Thus the majority of the people do not give any attention to these policies and tend to produce, exploit and rationally use their own energy resources as long as there is no objection from anyone. The relevant authorities seem to support this position by passively handling energy policy matters and doing almost nothing to diversify energy resources to increase options to the people.

In addition to health and environmental protection, the policies also cover gender and use of public forests. The policies specifically recognize the role of women in household energy management and seeks to improve their participation in energy development programmes and open to them ownership of energy resources at both demand and supply ends. It is therefore expected that deliberate efforts are made to ensure that women participate in energy related education, training, planning, decision-making and implementation strategies but, in practice, this is not visible in energy management strategies. With regard to forest resources, the policies are concerned with sustainable exploitation without adverse impact on the ecosystems or reduction in biodiversity. All these policy provisions indicate that national energy concerns in the region only cover the following four issues:

1. Improved access to energy.
2. Health.
3. Protection of forests.
4. Elimination of gender disparity in energy issues.

The implications of these policies is that the use of various energy sources should not endanger the health and lives of people while keeping the environment, ecosystems and biodiversity protected from any form of degradation. The rights of women who are normally overburdened with the task of fetching biomass-based fuels are also recognized and it is expected that these would be protected through ownership rights.

It is important to trace the background of these policies in order to understand why the governments are cold about the development of renewable energies. Before 1979, there were no government departments with overall mandate on energy issues. Energy matters were widely dispersed in different government ministries and, as a result, there was total lack of policy guidelines on energy development. When relevant ministries or departments were eventually established, budgetary allocations to them were very low and apparently this has not changed even where full Energy Ministries have been established. It is for this reason that NGOs and the private sector in general became actively involved in the development of renewable energy technologies and their dissemination in East Africa. Consequently, renewable energy activities have followed an ad hoc path with very little input, if any, from national governments. Whenever they got involved in renewable energy activities, they did so because donors and the international community in general were interested in renewable energy applications, mainly for environmental reasons. The main energy concern was in oil business and electricity generation and when renewable energies began to appear in the policy documents, it was because there was too much talk about renewables in the international forums, but at the national level, nothing was done to attract investors to renewable energies. To some extent, the monopolistic practices of government-controlled power utilities and multinational oil companies have strangled the official interest in renewable energies. To demonstrate this, we look at the Electricity Acts, which were basically formulated under very close supervision of the national utility companies and naturally gave the same companies power and authority to control a wide range of energy issues. These Acts prohibited commercial generation and sale of electricity directly to the consumers but gave the same utility companies the mandate to issue licenses for commercial electricity generation. The electricity thus generated under license would be sold to the same utility company, more or less, at its own price. This arrangement that made the utility companies (Uganda Electricity Board, Kenya Power and Lighting Company and Tanzania Electricity Supply Company) the custodians of the Electricity Acts is the reason for lack of support for renewable energy development. The system made it extremely difficult to get the license and make profit out of electricity generation. Although these Acts have been reviewed, there are still many obstacles and constraints imposed through pricing and long bureaucratic processes for private sector investment on electricity generation. Thus the policy change and the inclusion of renewable energies in the policy documents is still not enough to address the entire energy situation. There is need to turn the situation round and provide commercial incentives for power generation, particularly for use of renewable energy resources such as solar, wind and small hydro schemes. National utility companies should be treated like any other service provider operating and obeying the provisions of the Act and not as the custodian of the Act, a facilitator and a competitor at the same time. Tanzanian government has made some bold policy steps to attract investors into the development of renewable energies by providing attractive financial terms for potential investors. It has simplified procedures for investing in solar, wind and micro hydro schemes including a 100% depreciation allowance in the first year of operation. There are also exemptions from excise duty and sales tax and concessionary customs duty on the first import of materials used in renewable energy projects. In addition, extensive guarantees are provided to investors under the investment promotion centers certificate of approval. There are other attractive terms on repatriation of income, ownership guarantees and dispensation of assets. These are all very positive steps towards the development of locally available renewable energy resources that should be emulated by the other states.

However there is still a lot to be done to remove the control roles of national utility companies involved in the generation, distribution and sale of energy and create an environment in which the policies can be implemented freely and effectively. As mentioned above, these policies, with the exception of a few initiatives, exist only as official records of intention but very little effort, if any, has been made to achieve them. Often, some of them are used to discourage potential new players. The fact that these policies came into being only recently also shows how passive the governments have been about the development of the energy sector. The question is whether the governments understand the reasons behind the on-going vibrant global energy politics and whether they are sensible enough to use the currently available resources to fully prepare for the imminent future energy shortages. A good policy, if implemented, can cushion some of the adverse impacts of global energy shortages. For the developing nations, this needs effective mobilization of both domestic and external resources to increase investment in the development of local energy resources by deliberately creating conducive business environment for renewable energy technologies. This will further require efficient support from established institutional arrangement in which energy initiatives receive due attention without unnecessary bureaucratic procedures. In order to provide meaningful accessibility to energy, the governments must provide correct microeconomic facilities to both energy producers and consumers through balanced energy pricing that would satisfy the needs and circumstances of both parties. Appropriate regulatory measures that would encourage people to adopt efficient energy production and clean energy use should be constituted and implemented at various levels. Development of local resources and financial incentives should form the basis of these initiatives. This will ensure that commercial energy companies, especially electricity generating and distribution companies, aggressively pursue network expansion policy as the only way of generating income and not through over-taxing a few consumers. However it is worth noting that, in the past, the partially government-owned utility companies received financial assistance from public funds but did not improve performance and coverage as was expected. Cases in which people paid for new line connections but had to wait for years to be connected were so common that consumers could not understand the business of such power companies. Such delays are today still very common despite the many promises and the new energy policies. This is yet another reason why it will be absolutely necessary to put more effort on policy implementation instead of spending time on policy formulations which are eventually not enforced. There is a need to improve on the implementation of existing policies, this is to say, implementation efficiency must come first and this must be done in an environment of competition and operational self-sufficiency without expecting special "survival kit" from the government. The consumer, on the other hand, must be the ultimate beneficiary of the regulations and incentives through appropriate and effective policy enforcement. Due to the governments' past interest and stake in the utility companies, the common practice was to protect the companies and not the consumers and as a result the electrification process progressed at an extremely slow rate.

The inefficiency and irresponsible management by government departments affected other resources as well. The development of biomass resources, which had all the requirements for success, could not be sustained because the very people who were expected to enforce the policy on protection of forests and biodiversity also abused it. In Kenya, for example, the government protected all forests on trust land by posting government-appointed forest officers to all forested areas to ensure that the surrounding communities adhered to the laid down procedures for harvesting forest resources. But these same officers turned round and irresponsibly sold forest resources for their own benefits. The policies are still in force but most forests are gone. Similarly the policies on energy and gender; energy and health; and energy accessibility are still very much alive but they are not implemented. The point here, is that, as much as it is very important to have progressive policies, it is equally important to effectively implement those policies in order to achieve the desired goals. The private sector and many policy analysts have in the past recommended aggressive long-term policies that demonstrate economic and environmental benefits of energy resources, particularly Renewable Energies, with emphasis on job creation and income generation but not much has been achieved. It would appear like most of these suggestions also died of the common disease like most of their predecessors: implementation deficiency.

The benefits of diversifying energy sources including long-term savings on oil imports do not make any sense to the rural poor peasant farmers whose energy needs are restricted to heat, which can be met by using biomass and rudimental devices. Their needs for services that require petroleum fuels are only occasional and so it is not easy for them to relate their lifestyle with electricity or oil. Therefore expecting them to participate in the implementation of policies, which do not lead them to viable alternative energies, is unrealistic. Thus energy policies that would broaden the choices must target the consumers and encourage energy use strategies that would provide both incentive and socio-economic empowerment. Consequently, poverty reduction, accessibility to energy, suitable pricing and appropriate incentives must be properly integrated in the policy framework and effectively implemented. One of the approaches would be to prepare and implement clear strategic programmes and regulations that encourage active participation of key stakeholders including a wide range of consumer categories such as church organizations, institutions, community-based organizations, NGOs etc. All areas that are impacted by energy use such as health, housing and forestry and those that affect energy development and use such as energy technologies, financing and community energy-use-culture must all be considered in the programmes. It is also necessary to recognize differences in zonal energy requirements and availability and make appropriate provisions for energy use that are specific to those regions and their energy endowments. Since poverty is a major draw back in diversifying energy choices, it would be necessary to include income generation, health and gender issues in all energy programmes. In this respect, education, capacity building and general awareness creation would form an important component of such initiatives. A good policy should encompass all key areas including but not limited to health, financing, safety, education, accessibility, incentives, good energy mix, buying and selling arrangements with independent power producers. But, most importantly, the policies must be aggressively implemented at all levels through well-planed energy development strategies.

A very dangerous trend that needs to be seriously considered is the continued reliance on firewood and charcoal. As population increases, the demand for land for food production and human settlement would substantially go up and inevitably there will be increased demand for firewood. There are several possible results of this; one is the increase in price of firewood, which will also have its own repercussions. Increased demand for firewood also implies increase in the level of poverty with dire social and economic consequences including insecurity. The firewood crisis may also adversely affect food supply as both food and energy production will compete for the little land available. It is therefore not wise to address the biomass energy problem by trying to increase its production to meet the demand. This option should be taken as a short-term measure but, in the long run, it would create a far bigger problem than can be imagined. There will be a limit to which available land can supply adequate biomass energy and every effort should be made to avoid getting to this limit and the only option is to consider developing other inexhaustible or more elastic energy resources to provide heat energy to reduce the demand on firewood. Thus, whereas it is important to plan for more biomass production for obvious reasons, it is also equally important to start looking for means of reducing pressure on biomass as an energy source. The use of raw biomass mainly as a source of heat energy is not only inefficient but also does not give the versatility that is needed from an energy source. In this regard, it is necessary to also consider technologies that would transform biomass into a more versatile energy source such as liquid or gaseous fuel, which can produce energy for both heat and lighting and provide energy to run machines as well. Attention should be turned to fuels like bio-diesel and co-generation technologies that effectively broaden the energy base. Use of municipal wastes for electricity generation should be given serious considerations. Such developments will encourage use of clean energy and stimulate growth of new urban settlements, which, in turn, may ease the pressure on agricultural land. Since there are numerous issues that need to be addressed by a progressive energy policy, it becomes necessary to identify and formulate them into broad energy policy objectives that can be achieved. The broad energy policy objective should be to ensure adequate and quality supply of energy at affordable cost in order to sufficiently meet development needs while protecting and conserving the environment. Specific objectives that can be implemented with periodically measurable achievements are then formulated around this broad objective. Some of these are to:

- Improve access to affordable energy services;
- Enhance security of supply;
- Develop indigenous and renewable energy resources;
- Develop and instil energy efficiency and conservation practices as well as prudent environmental, health and safety measures;
- Develop a broad spectrum of local energy options to facilitate optimum mix in various applications.

Any serious energy development strategy must first recognize the crucial role of energy as a tool for socio-economic and industrial developments so that energy analysis becomes part of every initiative. There are therefore a number of challenges in the implementation of energy policy. Some of these include but not limited to:

- Establishment of effective legal, regulatory and institutional frameworks for creating both consumer and investor friendly environment;
- Expansion and up-grading of energy infrastructure;
- Enhancing and achieving economic competitiveness and efficiency in energy production, supply and delivery;
- Mobilization of resources for the required improvements in the energy sector.

In addition to these broad policy implications and challenges, there are issues that are pertinent to specific energy forms and sources such as electricity, petroleum and renewable sub-sectors. These must also be addressed in the context of overall energy objectives for both medium and long-term strategies.

There are also institutional and management challenges that have affected the expansion of the power sector in the region but the single most crucial one is the persistent inability to establish management and technological framework for the development of non-hydro energy resources. As discussed else where in this book, the power sector reforms that were introduced in the 1990s have not produced the desired results. These reforms were mostly concerned about the removal of the monopoly of national power companies such as KPLC of Kenya, TANESCO of Tanzania and UEB of Uganda that controlled the generation, transmission and distribution of electricity in the respective countries. It is evidently clear that the removal of this monopoly by creating additional companies has neither encouraged diversification nor spurred the expansion of generation capacity. It was however hoped that the move would reduce the impact of rampant mismanagement in the power sector. This did not happen and there is a growing concern about the management structures of the power companies that accept political interference. The Tanzanian government attempted to address this problem by contacting a South African company to run TANESCO, but the new managers faced internal resistance by local workers. At the end of the first two-year contract, a number of senior government officers were not convinced that the new managers improved the services. Despite the doubt, the contract was again renewed for another two years. Kenya is also likely to follow the Tanzanian policy since it is seriously considering contracting a European company to manage KPLC on a two-year experimental basis. It is understood that the company under consideration has extensive experience in managing similar power companies in several countries in Africa and Eastern Europe. The challenge however, is whether the expatriates will be fully accepted by the KPLC workforce, change the mind-set and operational cultures that have been established over the years.

As much as there may be management problems in the utility companies, the solution to this may not be found in the management style but in the ability to diversify electricity generation. In fact, many industries have made great effort to persuade the governments to be more serious and supportive on this front. For example, Tanzanian industries have pushed for suitable incentives to be established for electricity generation using non-hydro energy resources in order to avoid frequent power shedding due to low water level in the hydro dams. Uganda government on the other hand is responding to the low power output due to drought by planning to procure about 100MW of emergency thermal power supply. It is worth noting that Uganda's hydro power output relies on the level on water in Lake Victoria while has significantly gone down as a result of the 2005/2006 drought. Consequently Uganda's power generation has substantially decreased and this has led to serious power rationing in the country. The impact of the drought is even more severe in Tanzania where hydropower generation level drastically dropped from about 560MW to an all-time low of about 50MW at the end of February 2006. The general decrease of power output in the region has forced many industries to close down their operations.

7.8 Concluding Remarks

An outline of the available options and challenges regarding the development and commercialization of different energy resources is given in this chapter. Supportive information is obtainable from the following references [1, 17, 18, 22, 23, 24, 28, 29, 32, 37, 39, 41, 43, 44, 45, 49, 51, 52, 60].

8

The Way Forward

In this section, some suggestions are presented as guidelines to energy development in East Africa. However, they should not be treated as the only road map for improving energy situation in the region. There are unique circumstances that may require completely different approaches. Many proposals presented in the various chapters of this book should also be considered alongside these guidelines. It is now clear that there are a number of players in energy development and so the planning and implementation strategies must involve the various players and not the ministries concerned alone. In fact in energy matters, everyone is a stakeholder particularly as consumers. The only cause for variations in the methods and types of energy applications is that some stakeholders are rich while others are poor and this causes remarkable differences in the pattern of end-use energy mix. There are many factors influencing energy availability and methods of use and the associated problems that need to be addressed are also highly variable covering a wide range of social, economic, environmental and technical issues that are so much woven together that it becomes impossible to achieve anything by approaching them one at a time. They require a comprehensive multi-sectoral approach that is well planned to fully benefit from both local and international resources with clear benchmark for private sector participation and collaboration with national governments on all aspects while at the same time giving international development partners clear priority areas for support. But first, at the local level, multi-sectoral energy needs must be properly and accurately assessed with rural energy supply as a special case. It is in deed special due to the conditions of shelters, poverty and the scattered nature of the households. Proper documentations on various energy potentials, precise locations and preferred development strategies as well as the level of required investments should be compiled and made easily accessible to all players. As is now clear, all the three East African states have some rural energy plans but lack the institutional framework for effective implementation. For example, Uganda has Energy for Rural Transformation (ERR) programme with set targets for raising rural electricity access level; Tanzania and Kenya also have some programmes with similar objectives. But all these have not specified how the resources and experience in the private sector including NGOs are going to be tapped or coordinated to augment governments' efforts in areas such as technology transfer, financing, subsidies, investment priorities and incentives. At the international level, there are no guidelines for attracting donor support and utilizing resources from bodies such as GEE and other environmental related international protocols that provide support for reductions in emissions of greenhouse gases. There is need to design specific programmes in order to access the available funds for related sectoral projects. For rural energy programmes, affordability and access are important factors that should not be left out of the plans. To complete the foundation for future energy development, the on-going energy sector reforms should be re-examined. Some aspects of these have had negative impacts on poverty reduction efforts while others may be in conflict with the energy development objectives of the national governments. Examples include increased tariffs and missed opportunities for attracting investment on off-grid power production. If these are not checked then more and more people are likely to turn to biomass use despite the desire to move to cleaner energies. This would create long term problems since, even in the biomass situation, the governments do not have active policies and budgetary interventions to achieve sustainable biomass use. National energy policies should be focused on the beneficiaries' priorities of which affordability, security of supply and safety of applications are the most important. But in all these, poverty will remain a major drawback in any initiative that aims at supplying clean energy to the people and so both poverty and energy issues must be addressed in relation to one another. Consideration of all these as well as institutional framework to deal with them should form the basis for the implementation of energy policies. Having said these, it is important to recognize the efforts that have been or are being made by individual East African governments. Uganda government, for example, has made great efforts to ensure that private sector is attracted to the energy sector, especially in the rural electrification programme where renewable energy sources are expected to play a key role under Renewable Energy Board. At the same time it is working with the private sector to provide some financing arrangements and subsidies so that the poor rural communities can obtain renewable energy appliances such as solar panels and their accessories under easy payment terms. The other East African states should also fine-tune their policies and plans on renewable energies.

Considering national energy concerns, it is clear that they have been associated more with industrial operations than with rural or general domestic services. Furthermore, energy has not been considered as a factor in poverty eradication strategies. This attitude has relegated the importance of energy to a very low priority level and therefore energy receives very little budgetary attention from the national governments despite the fact that it is the driving force in all development processes at all levels and in all sectors. Consider the situation in the Lake Victoria region of East Africa as an example. Fish is a very valuable product not only for the people living in the lake's basin but also for the economies of the national governments in general. Globally, fresh water fish is a popular delicacy and so fish from Lake Victoria is exported to Europe, Middle East and as far as Japan. The annual production from lake Victoria is estimated to be about 5,000,000 tonnes with local value of about 75 million US dollars. The export value is about ten times greater than this. However, the local fishermen have remained poor for a number of reasons despite the high export value of fish. There are two energy-related reasons for this. First, fishing is generally done at night in small boats and therefore fishermen have to use both kerosene and diesel (or gasoline) for lighting and running boat engines respectively. To catch the popular lake Victoria sardines, they have to be attracted to one area where they can be scooped out in large numbers since catching them one by one would not make sense because they are too small; mature sardine is less than 3 cm in length. Attracting them to a selected area can only be done at night using bright kerosene lanterns placed on a floating structure. The fishermen, sitting quietly in a canoe, would wait until there are a large number of sardines under the lantern and then quickly surround them with the net and scoop them out into the boat. It is believed that the sardines eat some of the night flying insects that fall into the water, having been attracted by the bright lights. This method is labour-intensive and requires additional costly inputs in the form of stable canoes, outboard engines, kerosene, lanterns and nets. There are over 50,000 small canoes involved in this type of fishing on the Kenyan side of the lake alone and it is estimated that they spend a total of about US dollars one million worth of kerosene per month. This comes to about twelve million US dollars per year and, of course, there are other costs that are not included in this estimate. A fraction of the fish catch (about 30%) that is not exported, is smoked or sun-dried in order to prolong their shelf life and make it possible to transport to long distant markets to fetch better price. This also requires additional energy input, which for the Kenyan portion of Lake Victoria is obtained from about 20 tonnes of firewood per month. When all other opportunity costs are included, then the total cost of fishing becomes too high at over 50% of the local market value. The second reason for poverty among fishermen is that they do not have access to electricity to facilitate preservation of fish and therefore they are forced to sell it almost at the price of the buyer in order to avoid total loss due to the rapid deterioration of fish under the hot equatorial climate. Despite the valuable foreign exchange that fish export earns, the governments of the East African states have not adequately addressed the energy problems in the fisheries and this has severely aggravated the level of poverty among fisher folks. This is but one example in which lack of energy translates directly into desperation and poverty.

There is no doubt that biomass is a major source of energy in East Africa and the general policy objective has been to ensure sustainable wood supply while avoiding environmental degradation at the same time. This was to be done through promotion of production of fast-growing trees as part of on-farm agricultural activities. Along with this, it was expected that various NGOs would support promotion of more efficient stoves and charcoal production methods. Indeed there were times when these activities seemed to be heading for full scale success particularly in Kenya where extensive tree planting and new stove developments were vigorously conducted. However, with population increase, diminishing household land sizes, unpredictable weather pattern and widespread poverty, biomass regeneration has drastically declined while environmental degradation has increased in the rural areas and negatively impacted on food security. The degrees of these events differ from country to country within the region. Kenya, for example, has experienced very interesting developments that are in contrast to those in Uganda and Tanzania. While Kenya's industrial development appears to be ahead of both Uganda and Tanzania, the level of general poverty of the people has been increasing. Presently those who live below poverty line in Kenya constitute about 56% of the total population while those of Uganda and Tanzania are below 40% of their total population respectively. But they also have lower per capita energy consumption than Kenya, a situation that seems to negate energy consumption level as an indicator of the level of development. A more developed country should have lower level of poverty. This notwithstanding, it means that both Uganda and Tanzania will find it easier to switch from traditional rudimental biomass energy use to cleaner commercial energies like electricity and LPG, if affordability is the only factor. However, there are a number of factors that together determine the choice of energy as mentioned in the earlier chapters. Energy policy must therefore be integrated into various aspects and levels of socio-economic development and should deal with the peculiarities of each source more specifically.

East Africa is a non oil-producing region and totally depends on imported petroleum. Therefore liberalization policy on procurement, distribution and pricing of petroleum products cannot ensure its steady supply all the time. Any instability in the oil producing countries or ripples in the international oil market, caused by the many global economic factors, can severely affect oil supply and prices within the local regional market. Such impacts have been repeatedly witnessed in the past. Furthermore, distribution of petroleum fuels is understandably more favourable to the urban areas that normally have several distribution points within easy reach of most consumers. In the rural areas, availability is not always guaranteed because the few retailers who deal in such products keep only small quantities at a time and there is very little that can be done about this in terms of government policy. However, the governments, through long-term planning, can begin to invest in the transport sector, which consumes most of the imported oil by supporting studies on the development of non-motorized vehicles and solar car. The solar car project is already picking up in some industrialized countries such as Japan, USA, and a number of European countries. In September 2005, twenty two solar cars took part in the eighth World Solar challenge held in Australia. The 22 cars with average speeds of about 90 kilometers per hour, raced through Australia from Darwin to Adelaide, a total of 3,000 km. This event was not just a challenge to the car manufacturers but also an encouragement to the developing countries located in the tropics where these is abundant sunshine. East Africa is right on the equator and could benefit from such developments and should get involved in such initiatives right from the early stages.

Electricity, on the other hand, is generated from large centralized facilities most of which are controlled by the national governments. In the past, the generation, transmission and distribution arrangements did not provide the required incentives for network expansion. Many aspects of the policies on these seemed to have been deliberately designed to discourage demand for power line expansion. This view is supported by the fact that electricity generation capacity is limited and therefore demand must be kept within the power generation growth limit. This is probably the reason why the power companies which were themselves struggling to make profit were in charge of practically all electricity matters including rural electrification programmes. These responsibilities enabled the utility companies to stifle demand and keep grid network expansion within the existing generation capacity. It was therefore not reasonable to expect them to spearhead rural electrification programme and achieve anything without massive expansion of the installed capacity. So despite financial support from the national governments, they achieved very little as has been discussed elsewhere in this book. Electricity pricing itself is a disincentive to many consumers who are aware of the extra charges that they are to pay over and above the consumed power. In Kenya, for example, as low as 30% of the consumer's electricity bill can be for the consumed power units but this depends on the power consumption level. The rest is composed of taxes and levies which are hardly re-invested into the sub-sector to improve supply. This is particularly so for small consumers whose monthly average bill is below USD 6.00. For the example of a bill shown in Table 8.1[1], 52% of the bill is for the consumed power units while 48% is composed of taxes and levies.

This percentage decreases with lower consumptions and may drop as low as 30%. As an illustration, if in the above sample bill the consumption dropped to 50 units, then the total payable amount would drop to about USD 3.50and the cost of the consumed units would be

---

[1]Note: The bill is for a conservative small household consumption using two or three lighting points and a small low power radio. Kenya Shillings 532.25 in the year 2000 was equivalent to US dollars 7.30 and the net monthly salary for this group is in the order of USD 100.

Table 8.1. Typical monthly house electricity bill for low-income family in Kenya (2000; in Kesh)

Item	Kesh.
Monthly fixed charge	75.00
Consumption (80 kwhr)	277.00
Fuel cost adjustment (164 cts/unit)	131.20
Forel adjustment (41 cts/unit)	32.80
Electricity Regulatory Board (ERR) Levy (3cts/kwh)	2.40
Rural Electrification Programme (REP) Levy (5%)	13.85
VAT (18%) - (200 units exempted)	–
Total bill payable	532.25

only 30% of the total amount payable to the utility company. For a consumption of 70 kwhr the actual cost of consumed units would drop to about 48% of the total while higher consumption units would constitute a bigger fraction of the bill and may be as high as 70%. In terms of the bill composition, a poor family that consumes very little electricity pays relatively more taxes than a rich family that consumes more power. These variations apply to consumptions under the same tariff since monthly fixed charge depends on the tariff, which is based on estimated consumption level. In general monthly fixed charges range from USD 1.00 for a small domestic consumer to about USD 100 for commercial and industrial consumers. Obviously these levies and taxes are high but many people would not mind if they were used to improve electricity supply to the needy. It would also make more sense if similar levies and taxes were imposed on imported fuels for the purposes of developing local energy resources. Tanzania, which appears to be more sympathetic to her poor citizens, introduced pre-paid electricity meters (LUKE) in 1995 to enable consumers to decide how much to spend on electricity. This gave the consumers the responsibility of planning energy consumption and adopting energy mix that would ensure affordable overall expenditure on household energy. As a result, many consumers immediately learned the importance of energy conservation because of the obvious direct consequences of irrational use of electricity and began to demand more efficient (power-saving) lights. Such positive policies not only encourage consumers to have access to electricity but also raise their awareness on energy conservation. Even the very poor rural communities would want to use electricity whenever they are in a position to pay for it. Considering these two different approaches to electricity charges, it is clear that the Kenyan policy is retrogressive and has had negative impact on rural electrification. Whereas there may be good reasons for such a policy, it would be better to explain them to the people so that they do not feel disadvantaged and exploited by the policy.

In order to find a good energy mix for the rural communities, it is perhaps more logical to look at the energy end-use and then try and find a suitable source that can supply that energy and then think of a good policy that can facilitate the process. For example, rather than consider the problems associated with biomass energy, it would be better to consider 'heat' as the energy for cooking and formulate a policy that would examine various possibilities of heat production. Similarly consideration can be given to 'transportation', 'lighting' and 'machine power'. Although this is not a familiar approach to energy issues, it is nevertheless necessary in a complex situation where poverty and living conditions limit the choices. Thus the authorities in East Africa must, first and foremost, recognize the importance of energy in development of any country and also appreciate the magnitude of the barriers to the promotion of the locally available energy resources particularly renewable energies and the efforts needed to address them. The challenges therefore lie in mobilizing financial and human resources as well as establishment of incentives for the private sector to invest in technology innovations that would expand the market by diversifying supply sources and mechanisms. The potentials and possibilities for achieving all these exist and the onus is on the national governments to create enabling environment for active and well-coordinated renewable energy investments. An attempt will be made to make recommendations along this line.

8.1 Some Guidelines

To effectively address energy problems in the developing countries, it is important, first, to recognize the large difference that exists between commercial energy use and domestic energy requirements because this is the reason why different sources are used for their provisions. The heavy reliance on biomass for domestic energy is a major drawback in that this form of energy cannot be used to increase the level of income for the people making it difficult to raise the standard of living. Other sources of modern energy such as nuclear, geothermal, large hydro and fossil-based fuels are not only too expensive for the rural poor and developing countries in general but are also unsuitable for the conditions of the majority of the people. For many years, the sun and biomass have been the sources of energies that provide environmental comfort for most people in the rural areas of developing countries while animals and some oil-powered machines provide mechanical power. Kerosene also provides energy for cooking and lighting but this is limited by distribution problems and high operational costs. Although the situation is not likely to significantly change in the near future, it is important to increase energy base by finding alternatives to the rapidly decreasing wood supplies. However, there should be much more realism in the formation of programmes for alternative renewable energies. Given the magnitude and the unique nature of domestic energy production, distribution and end-use requirements of the rural population, the guidelines presented in this section focus on possible sources of forms of energy required for domestic application and the mix that can eventually lead to use of modern clean energy sources. Therefore the recommendations given in this section are based on the fundamentals of rural energy problems that are concerned about lighting and high temperature heat energies.

Heat is the basic form of energy required by all households in both urban and rural areas. It is used principally for cooking but it also serves other less essential requirements such as space heating. Basically all energy sources can supply heat but many of them cannot achieve the high temperatures at the rate required for cooking without using complicated conversion appliances. For example solar thermal energy when harnessed using simple technologies cannot achieve cooking temperatures at the required rate and this heat is not available all the time. It is therefore both inadequate and unreliable for the intended purpose and therefore nobody would want to depend on such energy. On the other hand, high voltage electricity from generating facilities can provide enough heat energy for domestic applications but provision of heat through this method is expensive and the appliances for its use are also expensive. Other energy sources and their conversion technologies can provide heat by first using them to generate electricity which has the same application constraints as discussed above. Wind energy conversion machines are also too expensive for most people to afford and there would also be need to invest in energy storage facility since wind energy, like solar, is intermittent. Similarly, other sources of heat such as oil and coal are expensive when used solely for heat production and they are not equitably distributed around the world. So the poor citizens of the developing countries whose average monthly income is less than USD 50 cannot afford heat energy from these sources. By directly burning biomass materials, sufficient heat can be produced for cooking and general heating purposes using cheap rudimentary devices. It is therefore evidently clear that as long as the economies of East African states are unable to significantly reduce the level of poverty to a small fraction of the total population, large quantities of biomass will continue to be consumed as the main source of heat energy for the majority of households. Poverty reduction is a very slow process and therefore, at present, more attention should be given to biomass as a source of heat. To strengthen this assertion, let us assume that the governments are willing to make high voltage electricity accessible to most people at the price of charcoal. Under the present crowded urban slums and poor conditions of rural shelters especially grass thatched mud houses, high voltage electricity would pose grave dangers to the people. This is one reason why it is important to reduce poverty to a level when people will be able to afford shelters that are suitable for modern clean energy use. So to save vegetatious and prevent land degradation, poverty must be squarely addressed.

In East Africa, only Uganda has established an elaborate Poverty Eradication Action Plan (PEAP), which gives good examples of the pivotal role electricity plays in improving the welfare and living standards of the poor rural population. For the people to benefit from electricity, they must be able to afford it in terms of cost and suitable application environment. Therefore both poverty eradication and energy provision must be addressed concurrently since the presence of electricity empowers people to reduce poverty level while the prevalence of poverty weakens the ability of the people to benefit from electricity. It is in this context that Uganda's PEAP is committed to reducing poverty from about 36% to less than 10% in the year 2017 while, at the same time, the government is providing enabling environment and incentives for private sector investment in electricity production specifically for rural electrification. Special programmes such as Energy for Rural Transformation (ERR) have been established in order to implement Rural Electrification Strategy. Some of the areas to be addressed in this process include power sector reforms, national policies that favour electricity production and rural electrification. It is only by effectively implementing such plans that the poor population can raise their income levels and improve quality of life. While Uganda has recognized the contribution and suitability of decentralized electricity generating facilities in this process and formulated detailed implementation programmes, the other sister states in East Africa that, basically, have similar policies on energy and poverty eradication, have not established specific programmes for achieving the goals. As a result, relatively many independent electricity producers have shown more interest in investing in Uganda than in either Kenya or Tanzania (see Fig.10 and note that Uganda is small in area and population compared to either Kenya or Tanzania). Uganda's approach is laudable. However, in the meantime, intensifying biomass regeneration activities should also be supported in order to minimize environmental degradation. Measures must therefore be put in place now to ensure that biomass production objectives are achieved, as people will continue to use it for cooking since food preparation is so crucial to people that they will do anything to get the energy for it. Preference would therefore be given to any energy mix that will guarantee security of supply at affordable price and application technology that is convenient to the rural circumstances. Let us look at the current energy mixes practiced by the various classes of people in general terms. Affluent households (about 5% of the total population) use various proportionate combinations of electricity and LPG while the middle and low-income groups use LPG/kerosene/charcoal and kerosene/charcoal/firewood respectively. About 80% of the population of East Africa is in the last category of energy mix and this is where attention should be focused. As much as there is need to generate more woody biomass, it is also necessary to reduce its consumption by moving more towards cleaner heat energy mixes through both economic empowerment and energy base expansion in terms of quality, availability and efficiency of application. The path that could be taken in this move is first to gradually shift from use of raw biomass to processed biomass energy resources such as charcoal, which can be obtained from a variety of biomass materials such as coffee husks, rice husks, saw dust, maize cobs and briquettes from other biomass wastes. In the raw form, many of these are not suitable sources of heat energy for cooking. However, their conversion into processed biomass energy sources will require a well-organized commercial fuel production system that can discourage haphazard and inefficient artisanal productions systems. Special institutes of energy development, if and when established, should focus on heat energy conversion and application techniques with a view to providing more efficient uses of biomass materials with guaranteed supply. It is only when this is done and made part of public energy domain that a plan for demand and supply management can be enhanced. In the meantime, and has been suggested, charcoal production from various biomass wastes should be raised to a commercial or industrial production level so that efficiency of production using well designed kilns can be enforced. This will also lead to high quality biomass fuels. It is not easy to improve wood-to-charcoal conversion efficiency under artisanal individual based systems. At present, most charcoal producers simply dig shallow pit kilns whenever they find enough wood to produce the quantity of charcoal desired. Training people to construct and use efficient kilns under such circumstances may not be effective since the kilns would be so much under-utilized that potential charcoal producers would not appreciate their contribution and therefore would not invest in them. On the other hand, significant achievement can be made at corporate production level involving cooperatives or recognized companies that have relatively long-term charcoal production plans. The East African Tanning and Extracting Company (EATEN) and Tanganyika Wattle Company (TANWAT) are good examples of successful use of improved kilns with efficiencies that are as high as 30% above the traditional kilns. Such initiatives need more support than they have received. Training individual charcoal producers to use efficient kilns may not yield significant results since such people often switch from one activity to another.

The next area of focus should be on improvement of efficiencies of charcoal and wood stoves. Some previous work, especially in Kenya, produced efficient stoves namely Kenya Ceramic Jiko, Kuni Mbili, Maendeleo and a variety of improved large institutional stoves. The emergence and success of these initiatives are attributed to the private sector that invested resources in the development and dissemination processes. National governments should come out and aggressively support dissemination of these stoves while at the same time making sure that quality stoves are produced and released into the market. In areas where farmers practice zero grazing of livestock and poultry keeping, household biogas production should be promoted. If possible, national veterinary departments should give some incentives to encourage farmers to invest in biogas plants. Biogas technology and application cannot be expected to naturally take root and establish itself because the cost involved is, in many cases, prohibitive. Deliberate efforts with some kind of incentives should be initiated with a view to broadening heat energy mix base for the rural communities. Apart from biomass-based fuels as sources of heat energy, the other heat energy mix at the bottom of the energy ladder is kerosene. Its use for heat is not as widespread as its application for lighting but a small number of rural households are increasingly getting into the practice of using kerosene for preparing quick meals that require only brief heating. School teachers, government extension officers and rural business communities are particularly attracted to using kerosene especially in the mornings when they have to quickly leave home for work. Local research should be conducted to develop safe and more efficient kerosene stoves at affordable price. Improvements on both charcoal and kerosene application devices and effective dissemination strategies may also encourage the emergence of a new heat energy mix that may group kerosene, charcoal and LPG together.

The next energy priority is for lighting, which is dominated by kerosene at the lower end. Other sources are biomass fuels, dry cells, lead-acid batteries and LPG but their contributions are very small compared to kerosene. Dry cells are usually used as flashlights for brief moments only. Cooking with firewood provides some lighting around the fire but is generally not used specifically to provide lighting. Table 8.2 gives an analysis of lighting costs from various sources of energy when used for up to four hours daily. For calculation purposes, the lifetime of grid electricity is fixed at 50 years with an initial investment of USD.600 (connection charges). This does not include the cost of wiring and elaborate switching systems in the house, which are normally required by the utility companies. It is clear that all fuels considered in this analysis including the popular kerosene are more expensive than grid electricity and solar PV systems. However, poor people still choose to use some of these expensive fuels such as kerosene and dry cells simply because the initial investment on the devices are low and it is possible to get them in small quantities. The costs quoted in Table 8.2 are based on the use of power-saving lights with a total of about 90 W for the household.

The cost of solar lighting systems and grid electricity are comparable but both are associated with high initial investment. The daily cost for solar PV includes battery replacements at least three times during the lifetime. The costs of appliances and grid power tariffs are assumed to be constant during the considered period. Under these assumptions, solar lighting has more advantages for rural household conditions than

Table 8.2. Daily lighting costs from various energy sources in US Dollars

Hours	Kerosene	LPG	Dry cells	Lead acid Batteries	Solar PV	Grid Power
1	0.06	0.15	0.80	0.06	-	0.04
2	0.12	0.30	1.40	0.12	-	0.05
3	0.18	0.45	2.01	0.18	-	0.06
One-time investment on device	6.50	21.00	1.14	71.4	300	600
Average life time of device in years	10	5	2	4	20	50
Daily cost on lifetime basis	0.18	0.46	2.10	0.23	0.07	0.10

grid electricity because of its low voltage that is not as dangerous to the user as high voltage grid power particularly under the poor and temporary conditions of rural shelters. Thus the best energy mix for lighting in the rural areas would be kerosene and solar PV systems. It is therefore important that rural lighting efforts be focused more on solar PV systems since kerosene lanterns are already in common use. There is, however, need to consider improvements of the efficiency of kerosene lanterns and production of cheaper versions. The on-going attempts to develop hurricane lamps that can use cheaper windshields than the traditional glass shields should be supported while at the same time encouraging production of new designs and various sizes of the lamps. On solar lighting systems, there are a number of low-cost new innovations that are exclusively addressing rural lighting problems. The version that would be suitable for rural households is a portable type in which the rechargeable battery and the lighting points are together in one unit. When in use, it can be conveniently carried from one room to another - just like the usual use of kerosene lamps in a rural household. Recharging of the unit takes place during the day in a convenient and safe location. Such low-cost solar lanterns should be readily available in the local markets and, if possible, under easy purchasing terms. It is also important to educate the target group on the limitations of such lanterns especially on daily maximum duration of use and minimum daily recharging time. This is necessary to ensure that user confidence is not eroded by disappointments due to uncontrolled use. Such lanterns can also be used in the fishing of Lake Victoria sardine that is generally attracted to the net at night using kerosene lanterns. The new method would definitely eliminate large expenditures on kerosene and improve the benefits to the fisher folks. But it might also encourage the use of kerosene as a source of heat energy in which case the pressure on fuel wood would be reduced, creating a balanced rural energy mix involving Kerosene, biomass and solar. A good rural energy policy should seek to establish this energy mix by improving security of supply of both kerosene and solar lanterns and associated efficient and affordable appliances. Large-scale use of solar lanterns will instill in people the culture of energy conservation and hence prepare them for the clean and versatile modern energy: grid electricity. The approach will also create the demand for grid electricity that is needed to spur the development of cottage industry and improve rural economic growth. However, for more organized cottage industry, decentralized energy supply facilities including small hydro, solar PV systems, wind and co-generation systems should be promoted to also serve as nucleation centers for rural electrification programmes. Where possible, like in Tanzania, natural gas and limited coal should be used to set up isolated electricity generating facilities specifically for rural power supply. Some of the resources that various ministries in charge of energy have used for consultancies and policy formulation should have been spent on support and establishment of national institutions for energy development in order to effectively address energy supply issues. The establishment of, for example, a regional or national energy development institute(s) is long overdue in the region.

Along with these developments, government departments in charge of agriculture, natural resources and forestry should work together with the private sector to systematically regenerate woody biomass as an integral part of agriculture and as special commercial biomass energy production particularly on government-owned land. This would encourage the emergence of commercial charcoal production systems that can benefit from improved kiln initiatives. The development of gaseous and liquid forms of biomass fuels should be sufficiently supported with a view to using them to blend imported liquid and gaseous fuels in order to reduce national bills on imports of such fuels. In this regard, attention must be focused on the development of bio-diesel and power alcohol. These are some of the technically viable biomass energy conversions that should be considered for adaptations and improvements in East Africa. The special energy institute referred to above should be responsible for research and developments in these areas while key stakeholders in petroleum and national grid electricity management review the pricing policies and introduction of suitable energy packaging for the consumers.

As mentioned earlier, good energy policies are of no use if they are not implemented. So it is necessary to establish institutional framework for specific policy implementations. Thus, in addition to the existing institutions of higher learning, there must be institutions responsible for energy research and development that can give guidelines on how to address both household and industrial energy needs without having to shift emphasis at the expense of the other. This is particularly important for rural electrification programmes, which, in the past, were the first to suffer discontinuation whenever there was any constraint in the energy sector. Such programmes should fall under an independent authority which should have very little to do with industrial energy management. And, to support energy research and development activities, there should be an institution that not only coordinates activities but also mobilizes funds for such activities. An establishment such as Energy Research Council would play this role. Where possible, part of the taxes levied on commercial fuels should be made available for local energy research and development. The general objective of such institutions would be to facilitate research, development and testing of new devices, prove the efficacy of energy devices and then set up functional units in some suitably selected locations. When all these arrangements have been put in place, attention should be turned to petroleum-based fuels. Recognizing the fact that oil supply to East Africa and its cost are sensitive to international demand-supply factors, measures must be put in place to ensure that, at any given time, the region has enough fuel supply to last several months, not just one or two months as is presently the case. This means that there is need not only to improve the refinery facility but also to expand the storage capacity and spread them in the region for both security reasons and ease of distribution. Expansion of pipeline network would achieve this objective and also reduce wear and tare on the road network. All these steps would require prudent use of national resources since oil imports require huge capital, which may not be readily available all the time and therefore adequate storage capacity is absolutely necessary in order to guard against unexpected disruptions due to local resource constraints or potential turbulence at the source.

Going back to the policy issues, it is instructive to note that although oil exploration has been going on in East Africa for decades, there has never been any discovery of oil deposit that is economically viable to exploit. So much energy and resources have been spent on this activity and yet very little is spent to develop what is already available like solar, wind, mini hydro and biomass potentials. The region has gone as far as establishing Petroleum Institute of East Africa when there is no deposit of petroleum resources in the region. On the other hand, there are abundant renewable energy resources such as solar, wind, small hydro and biomass. It would therefore be interesting to know why there is no Renewable Energy Institute of East Africa. It is very important for the national governments to get development priorities right in order to achieve meaningful progress and efficient use of locally available energy resources must be the nerve center of all initiatives.

Renewable energy has considerable potential in East Africa and should therefore occupy an important place in the region's energy programmes. Biomass, solar, hydro and wind power from small, decentralized units are capable of meeting a variety of energy needs in both the traditional and modern sectors. In this regard, the following recommendations, which are by no means exhaustive, should be considered by each country. In addition the unique cultural and socio-economic factors and renewable energy endowments should be taken into account.

i)   A comprehensive national energy policy covering production, supply, utilization and conservation should be developed and implemented by the individual countries in the region.
ii) Each country should have an extensive programme on local capacity building for renewable energy technologies.
iii) Special attention should be given to the development of small scale, decentralized generating facilities for all sources, particularly for rural electrification programmes.
iv) Research and development should focus on the production of equipment to harness renewable energies, instead of concentrating on extension work for imported equipment.
v)   The production of biomass materials as an energy source should be intensified.
vi) Research programmes to extract oil from crops as a direct substitute for fossil oil should be introduced.

These recommendations, if implemented, will speed up the development of renewable energy technologies in the region. Each country should, first and foremost, build a favourable, enabling environment in which the role of renewable energy can be defined well, prioritized among the various technologies and allocated adequate resources. In addition, specific policies and budgetary commitments must be made to facilitate the necessary coordination and cost-effective implementation of the programmes. This should include the extension and dissemination systems that are capable of reaching both the urban and rural populace, particularly the poor with technical and social assistance and credit facilities.

Some of the studies that have been conducted in the region have been very reluctant to recommend the development of certain renewable energy resources for the region despite the fact that some of the countries have serious and deep-rooted energy problems that can only be eased through the development of the renewable energy resources.

8.2 Concluding Remarks

There is a wide divergence of opinion about the policies that should be adopted in order to successfully develop suitable energy provision strategies for the generally poor rural communities. There are also issues concerning the required balance between domestic and industrial energy support or subsidies and also between commercial and non-commercial energies. The proposed guidelines for the way forward given in this chapter is but just one of the proposals that may work well for one region but may need modifications in order to succeed in another region. Such proposals and guidelines are available in [9, 10, 12, 17, 18, 23, 28, 29, 32, 37, 41, 46, 49, 51, 52, 60, 61, 64].

Appendix

Appendix: Energy Units and Conversion Factors

1 Calory (cal.) = 4.19
J 1 British Thermal Units (But) = 1060 J
1 Electron volt (eV) = 1.602 x 10-19 J
Energy Equivalences:

1 tonne of oil equivalent (TOE) equals:
42.2 GJ
0.93 tonnes gasoline
0.99 tonnes diesel oil
0.96 tonnes kerosene
1.04 tonnes fuel oil
0.93 tonnes LPG
1.61 tonnes coal
1.35 tonnes charcoal
2.63 tonnes fuel wood
6.25 tonnes bagasse
One barrel of oil is 0.159m3 and weighs about 136.4kg (density of oil
= 857.9kgm-3).
Power units:
1 Js-1 = 1 watt (W)
1 horse power (hp) = 746 W
1 But s-1 = 1060 W

Electricity consumption unit is kilowatt-hour (kwhr). One kwhr is
the quantity of power in kilowatts consumed in one hour. For exam-

ple, if a 50 W light bulb is used continuously for 20 hours, the total power consumption would be 1000 Watt-hours, which is equal to 1kWh. Thus using 1000W electric iron for one hour will consume 1 kwhr of electricity.

References

1. Anderson D (1986) Declining Tree Stocks in African Countries. World Development Vol.14, No 7
2. Anselm A (1981) Introduction to Semiconductor Theory. Mir Publishers, Moscow, Russia
3. Arunga RO, Nzomo MD (1988) Commerce and Industry. In Kenya: an official handbook. Government Printers, Nairobi, Kenya
4. Aseto O, Ong'ang'a O (2003) Lake Victoria (Kenya) and its environs: Resource, Opportunites and Challenges. Africa Herald Publishing House, Kendu Bay, Kenya.
5. Aseto O, Ong'ang'a O, Awange JL (2003) Poverty. A challenge for the Lake Victoria basin. OSIENALA Series 5, Printed by Africa Herald Publishing House, Kendu Bay, Kenya
6. Ashworth JH, Neuendorffer JW (1980) Matching Renewable Energy Systems to Village Level Energy Needs. SERI/TR 744-514, Solar Energy Research Institute, Colorado
7. Awange JL, Ong'ang'a O (2006) Lake Victoria: Ecology, Resources and Environment. Springer, Berlin Heidelberg
8. Bahati G (2003) Geothermal Energy in Uganda: country update. International Geothermal Conference, Reykjavik
9. Bhagavan MR, Karekezi S (1992) Energy for Rural Development. Zed Books Ltd, London, UK
10. Bosley K(1980) Large and Small Electric Wind Turbines for Isolated Application. In Twidell J(Ed.): Energy for Rural and Island Communities. Pergam Press, Oxford, UK.
11. Central Bureau of Statistics (2001) Economic Survey. Republic of Kenya
12. Curran SC, Curran JS (1979) Energy and Human Needs. Scottish Academic Press, Edinburgh, UK
13. De Renzo DJ (1979) Wind Power: Recent Developments. Noyes Data Corporation
14. East African Community (EAC) (1997) East African Cooperation Development Strategy, 1997–2000. Arusha Secretariate for the Permanent Tripartite Commission for the East African Cooperation
15. East African Community (EAC) (2004) Protocol for sustainable development of Lake Victoria basin. Published by the East African Community Secretariat

16. Economic Survey (2003) Economic Survey, 2003. Republic of Kenya, Central Bureau of Statistics, Ministry of Regional Planning and National Development, Nairobi, KENYA.
17. Flauret P (1978) Fuelwood Use in a Peasant Community: Tanzania case study. Journal of Development Areas
18. Goreau TJ, de Mello WZ (1988) Tropical Deforestation: some effects on atmospheric chemistry. Ambio 17:4, pp 275–181
19. Hall DO (1989) Carbon Flows in the Biosphere: present and future. Journal of Geological Society Vol. 146 pp. 175-181. London, UK
20. Hankins M (1989) Renewable Energy in Kenya. Motif Creative Arts Ltd, Nairobi Kenya
21. Hickman GM, Dickins WHG, Woods E (1973) The lands and poeple of East Africa. Longman, Essex, UK
22. Karekezi S, Ranja T (1997) Renewable Energy Technology in Africa. Zed Books Ltd, London, U.K
23. Karekezi S, Kimani J (2004) Have Power Sector Reforms Increased Access to Electricity Among the Poor in East Africa? The Journal of International Energy Initiatives, "Energy for Sustainable Development" vol VIII, No. 4
24. Kimani MJ, Naumann E (1993) Recent Experiences in Research, Development and Dissemination of Renewable Energy Technologies in Sub-Saharan Africa. KENGO International Outreach Department, Nairobi, Kenya
25. Kittel C (1996) Introduction to Solid Physics. John Wiley and Sons, New York, USA
26. Leaver KD, Chapman BN (1971) Thin Film. The Wykeham Science Series, London, UK
27. Levy RA (1972) Principles of Solid State Physics. Academic Press, New York, USA
28. Manibog FR (1984) Improved Cooking Stoves in Development Countries: Problems and opportunities. Annual Review of Energy, No. 9
29. Mapako M, Mbewe A (2004) Renewable and Energy for Rural Development in Sub-Saharan Africa. Zed Books Ltd, London, U.K
30. Monteith JL, Unsworth MN (1990) Principles of Environmental Physics, Edward Arnold, London, U.K
31. McNitt JR (1982) The Geothermal Potential in East Africa. Proceeding of the Regional Seminar on Geothermal Energy in Eastern and Southern Africa, Nairobi, Kenya
32. O'keefe P, Raskin F (1985) Fuelwood in Kenya. Ambio 14:4-5, pp 221-224
33. Okken PA, Swart RJ, Swerver, S (1989) Climate and Energy. Kluwer Academic Publishers, Dordrecht, The Netherlands
34. Oludhe C. Ph.D Thesis. Department of Meteorology, University of Nairobi
35. Ong'ang'a O (2002): Poverty and Wealth of fisherfolks in the Lake Victoria basin of Kenya. Africa Herald Publishing House, Kendu Bay, Kenya.
36. Ong'ang'a O, Othieno H, Munyirwa K (2001) Lake Victoria 2000 and Beyond. Challenges and opportunities OSIENALA, Kisumu
37. Onyebuchi EI (1989) Alternative Energy Strategies for Developing World's Domestic Use: a case study of Nigerian household's fuel use patterns and Preferences. Energy Journal Vol. 10, No. 3, pp 121-138
38. Othieno H (1985) Optimization of Solar Heating Collectors used for drop Drying. In Renewable Energy Development in Africa Vol.2 pp17-31 Commonwealth Science Council, London, UK

282

39. Othieno H (1988) Problems and Prospects of Technology Transfer. In Othieno H (Ed) Applications of Appropriate Technologies pp. 20-23

40. Othieno H, Franz E (1989) Report on the NGO conference and workshop on Prevention of Climate Change. Stichting Natuur en Milieu. Rotterdam, The Netherlands

41. Othieno H and Kapiyo R.J.A. (1989) Processes of Technology Transfer: some critical issues in developing countries. International Journal of Energy Exploration 7, 2:93-102

42. Othieno H (1990) Emerging Global Environmental Concerns. Proceedings of Environment 2000 Conference. UNEP, Nairobi, Kenya

43. Othieno H (1991) Alternative Energy Resources: the Kenya perspective. Energy and the Environment pp 413-415 ASHRAE, Atlanta, USA

44. Othieno H (1992) Alternative Energy Resources. Journal of Energy Sources Vol. 14:4 pp 405 -410

45. Othieno H (1993) Research and Development in Renewable Energy Technology in Sub-Saharan Africa. Renewable Energy Technologies in Sub-Saharan Africa. KENGO, Nairobi, Kenya

46. Othieno H (2000) Photovoltaic Technology: Most appropriate electricity source for rural tropical Africa. In Sayigh A.A.M (Ed) World Renewable Energy Congress VI (WREC2000). Elsevier Science Ltd. UK

47. Othieno H (2003). Principles of Applied Physics: Matter, Energy and Environment, One Touch Computers, Kisumu Kenya

48. Priest J (1984) Energy: Principles, problems, alternatives. Addison - Wesley, London,UK

49. Schipper L, Meyers (1992) Energy and Human Activity. Cambridge University Press, London, UK

50. Sexon B, Slack G, Musgrove P, Lipman N, Dunn P(1980) Aspects of Wind Energy Conversion System. In Twidell J(Ed.) "Energy for Rural and Island Communities". Pergam Press, Oxford, UK

51. Slesser M (1980) Energy for Remote Communities: the strategy. In Twidell J(Ed.) "Energy for Rural and Island Communities". Pergam Press, Oxford, UK

52. Smale TH (1980) Cogeneration of Heat and Power: a market opportunity. In Twidell J(Ed.) "Energy for Rural and Island Communities". Pergam Press, Oxford, UK

53. Sparknet (2004) Uganda Country Overview

54. Sparknet (2004) Tanzania Country Overview

55. Sparknet (2004) Kenya Country Overview

56. Swift DG (1983) Physics for Rural Development. John Wiley & Sons, Chichester UK

57. Thielheim KO (1982) Primary Energy: Present status and Future Perspectives. Springer - Verlag, New York, USA

58. Twidell J, Weir AD (1986) Renewable Energy Resources. E & F.N. Spon Ltd, London, U.K.

59. Uganda Bureau of Statistics (2001) Statistical Abstract

60. UNDP/World Bank (1984) Energy Issues and Options in Thirty Developing Countries. Report No.5230

61. UNDP (2002) Energy for Sustainable Development: a policy agenda. UNDP, New York, USA

62. UNEP (1979) The Environmental Impact of Production and use of Energy Part I, Fossil fuels. Energy report series 1979
63. UNEP (1986) The State of the Environment. Environment and Health. UNEP Report 1986
64. Van Lierop and Van Voldhuizen L.R (1982) Wind Energy Development in Kenya
65. Wilson JIB (1979) Solar Energy. The Wykeham Science Series, UK
66. Yu PY, Cardona M (1996). Fundamentals of Semiconductors. Springer, Berlin, Germany

Index